CREATIVE LIVES AND WORKS

The Science of Life

The Science of Life: Andrew Huxley, Richard Keynes and Horace Barlow falls under the series *Creative Lives and Works.* It is a collection of interviews conducted by one of England's leading social anthropologists and historians, Professor Alan Macfarlane. Filmed over a period of 40 years, the three conversations in this volume, are part of a larger set of interviews that cut across various disciplines, from the social sciences, the sciences, to the performing and visual arts. The current volume on two of England's foremost physiologists and a vision scientist is yet another addition to the series of several such books.

These Cambridge men of science, Sir Andrew Huxley, Richard Keynes and Horace Barlow, apart from shaping certain very fundamental and critical elements in the disciplines of Physiology and Neuroscience also belong to illustrious lineages. Sir Andrew Huxley, for instance is a direct descendant of T.H. Huxley, while Richard Keynes and Horace Barlow are both the great grandsons of Charles Darwin.

Their conversations greatly expand our understanding of physiology and neuroscience. The book will be of very great value not just to those interested in Physiology, Medicine and Neuroscience. The interviews also take us into a fascinating period of Cambridge Science, dominated by certain key families of distinguished thinkers.

Alan Macfarlane was born in Shillong, India, in 1941 and educated at the Dragon School, Sedbergh School, Oxford and London Universities where he received two Master's degrees and two doctorates. He is the author of over forty books, including *The Origins of English Individualism* (1978) and *Letters to Lily: On How the World Works* (2005). He has worked in England, Nepal, Japan and China as both an historian and anthropologist. He was elected to the British Academy in 1986 and is now Emeritus Professor of Anthropology at the University of Cambridge and a Life Fellow of King's College, Cambridge. Professor Macfarlane received the Huxley Memorial Medal, the highest honour of the Royal Anthropological Institute in 2012.

CREATIVE LIVES AND WORKS

The Science of Life
Andrew Huxley, Richard Keynes and Horace Barlow

In conversation with
Alan Macfarlane

Edited by
Radha Béteille

Routledge
Taylor & Francis Group
LONDON AND NEW YORK

SOCIAL
SCIENCE
PRESS

First published 2023
by Routledge
4 Park Square, Milton Park, Abingdon, Oxon OX14 4RN

and by Routledge
605 Third Avenue, New York, NY 10158

Routledge is an imprint of the Taylor & Francis Group, an informa business

British Library Cataloguing-in-Publication Data
A catalogue record for this book is available from the British Library

ISBN: 9781032405063 (hbk)
ISBN: 9781032405070 (pbk)
ISBN: 9781003353409 (ebk)

DOI: 10.4324/9781003353409

Typeset in Sabon Lt Std
by Manmohan Kumar, Delhi 110 035

SOCIAL
SCIENCE
PRESS

Contents

Preface vii
Note on Transcription – Radha Béteille xi
Introduction – Alan Macfarlane xiii

PART I

Andrew Fielding Huxley – In conversation with Alan Macfarlane 3

PART II

Richard Keynes – In conversation with Alan Macfarlane 39

PART III

Horace Barlow – In conversation with Alan Macfarlane 65

Appendix: Biographical information – Compiled by Radha Béteille 92

Preface

SIR ANDREW HUXLEY

I only met Sir Andrew twice, during the two interviews I carried out at his home just outside Cambridge. I remember being somewhat in awe of such a distinguished person – a Nobel Prize winner, the ex-Master of Trinity College, and grandson of T.H. Huxley (polymath and defender of Charles Darwin) one of the intellectual giants of the nineteenth century, and half-brother of the writer Aldous Huxley who I greatly admire. Yet when I met Sir Andrew, I found a very gentle, courteous and thoughtful man, not frightening or arrogant. The two interviews in his lovely house set in a beautiful garden in Grantchester, near where I had lived for a couple of years when I first came to Cambridge, were easy and relaxed.

RICHARD KEYNES

Keynes is another member of the illustrious Cambridge scientific families, the great-grandson of Charles Darwin, and related to others such as John Maynard Keynes the economist at my College (whose wardrobe I look after in King's). One of his sons, Simon, I knew through a shared interest in Anglo-Saxon history. I knew that Richard Keynes was the son of the very distinguished literary figure, Geoffrey Keynes, a biographer and expert on William Blake. So I was not surprised to find his large and lovely house on the Backs at Cambridge filled with family portraits, which he kindly showed me round. I always hoped to go back one day, but those days are past.

HORACE BARLOW

It is a coincidence, which I did not realize until I interviewed him, that Horace Barlow was also one of the Huxley – Darwin – Keynes clan of Cambridge. He was the great grand-son of Charles Darwin, one generation on from Sir Andrew Huxley. Like Huxley he was at the great science College, Trinity, and like Huxley made fundamental breakthroughs in neuro-science and particularly the science of vision and computation. I had been encouraged to interview him by my advisor in King's, Sir Gabriel Horn, and when I met Barlow I immediately discovered that I had met him walking round Cambridge on a number of occasions. Until the great influx of tourists in the last ten years, central Cambridge was really a village and one often encountered Nobel prize winners and others who had changed the world on the paths and in the courtyards. When Barlow, Keynes and Huxley were in their prime, it must have felt even more of a community of people who knew each other and worked together in the laboratories and libraries. This, as we can see from these three cases, was re-enforced not only by working in the same College and Department but also by family bonds, as described in some detail by Noel Annan in his book *The Dons* (1999) which contains a last chapter which analyzes these families and shows their connections to a certain form of agnostic, protestant, deistic belief.

Transcriber's Note

The transformation of videographed speeches into the written word has been a riveting experience. Transcriptions, while retaining the essence of a conversation, lend a new dimension to the oral transmission, opening them up to a wider readership, including those with diminished auditory functions.

This collection of transcripts lucidly captures the engaging interactions between Alan Macfarlane, one of Britain's foremost social anthropologists, and three eminent physiologists – Andrew Huxley, Horace Barlow and Richard Keynes. In fact, the intellectual discourses in these interviews on physiology and medicine draw in the reader as a participant. And, while offering a different order of internal cohesiveness, accentuates the readers' ability to retain information, in this case, knowledge and scholarship.

Here, it must however be noted that not being a subject expert has been somewhat limiting. Besides, I have had to acquaint myself with speech patterns, accents, diction and intonations in areas that I was unfamiliar with when I began.

Furthermore, the transcribing of oral transmission leaves little room to represent facial expressions and non-verbalized sounds such as chuckles, laughs, sighs and for that matter exclamations, as well as pauses and idiosyncrasies in actual speech. Thus, I have taken the liberty to replace them with closely corresponding punctuations, so as not to alter the meaning or import of what is being said. Discrepancies in speech patterns have been isolated and addressed. Any misrepresentation of words and phrases, is entirely inadvertent.

The trust and generosity of Professor Macfarlane in sharing his wealth of scholarship with Social Science Press cannot be adequately expressed in words.

– Radha Béteille

Introduction

Alan Macfarlane

There have been many autobiographical accounts of the creative process in science as well as in the arts, humanities and social sciences. Yet these tend to concentrate on one level, and within that one aspect, the cerebral, intellectual working of a single thinker's mind. If we are to investigate further the connections between the levels of civilization, institution, network and individual, and the fifth dimension of chance or random variation, we need to supplement these accounts, in particular by letting scientists and others talk in a relaxed way about what they think has been important in their lives and works. Over the years I have been collecting such data and here I would like to describe how this happened and what opportunities it opens up for further understanding of the springs of creativity.

***

There is a puzzle as to why I have spent so much time and effort (and expense) on interviewing (on film) a large number of academics and others over the years. No one else has done this as far as I am aware and it has few tangible rewards except the occasional gratitude of one's colleagues. Why and how did this project build up over the years to a point where I now have about 250 lengthy interviews, almost all of them on the web?

One factor is clearly my anthropological training and the experience of anthropological fieldwork. Although some social historians of the recent past began to become interested in oral history and the tape recording of memories from the late 1960s, for example

Paul Thompson, Brian Harrison and others, on the whole historians deal with dead people who cannot be interviewed. Certainly, as a 17th century historian I was not interested in interviews.

Anthropologists on the other hand deal with the living and the 'life history' or in-depth study of one or two 'informants' naturally leads into the use of recording devices – again usually tape recorders, but also sometimes film cameras, to record interviews. So, from my first fieldwork in 1968–70 in Nepal, I was aware of the potential of interviews. Alongside the interviews with intellectuals described below, I have made in-depth interviews with my anthropological subjects, in particular a ten-part interview with Dilmaya Gurung in 1992. I also taped and filmed members of my family talking about their lives, especially my grandmother in 1978.

My interest in biographical details as a source of social history was also stimulated by the accident of moving through a number of different intellectual landscapes in my early years. My grandmother's stories of Burma and India and then my mother's vivid autobiographical writings in her letters and when she visited me from India were early influences which led to an interest in listening to older people talking about their lives. So, when I became an undergraduate and then a graduate at Oxford (1960–66) I spent some time listening to elderly academics, particularly my history tutor Lady Rosalind Clay with her deep immersion in the world of Oxford life through her father A.L. Smith (Master of Balliol), Sir Henry Clay her husband and links to the Mitchison an Mitford families. She seemed to know personally many of the people whose books I was reading – Namier, Tawney, Trevelyan, Hill, Habakkuk, and she told me about their lives and personalities. I also began to hear about the folklore and gossip of historians from my other teachers – Harry Pitt, Keith Thomas and others.

The world of the historians was brought into relief by being on the edge of another group of academics, the social anthropologists. As I did my D.Phil. on witchcraft, I began to meet the older members of the distinguished school of anthropologists, the Lienhardts, John Beattie and above all E.E. Evans-Pritchard. This was a particularly intense and gossip-connected tribe and I heard their stories of the great feuds and friendships and how they had survived their fieldwork.

The interest in biography was also strengthened by my early apprenticeship in history at Oxford where I was taught how important it was to study not only the works but also the lives of great historians – Gibbon, Macaulay, Tocqueville, Bede and others. This led me much later in my life to write detailed studies of other major figures in books on Montesquieu, Adam Smith, Tocqueville, Maitland, Fukuzawa and others which put a strong emphasis on the biography as well as the ideas.

As a social historian I wanted to get inside the mentality as well as the social structure of the past. As an undergraduate I had discovered that letters (Pastons, Stonors) and diaries (Pepys, Kilvert) were wonderful sources for social history. So, my first book was a study of the life of one individual *The Family Life of Ralph Josselin*, based on the diary of a 17th century clergyman. Henceforth I collected letters, autobiographies, diaries, travellers' journals and made a great deal of use of this biographical material in a number of my books.

The early experience of eavesdropping on what appeared to be a disappearing world of a certain academic endeavour in the pre-war world of Oxford history and anthropology, was reinforced by moving to the London School of Economics for two years, where giants of the post-Malinowski generation, Raymond Firth, Isaac Schapera, Lucy Mair, were on the point of retiring. Then when I moved to the School of Oriental and African Studies (SOAS) I met another group of giants from a previous age, particularly Adrian Mayer and Christoph von Fürer-Haimendorf who became my supervisor.

When I moved as a research fellow to Cambridge in 1971, I encountered again several oral worlds which intrigued me. There were a new set of older anthropologists, Meyer Fortes, Audrey Richards, G.I. Jones, Edmund Leach, S.J. Tambiah, Jack Goody among them. And there were a group set of historians, in particular Peter Laslett and Tony Wrigley. A third world was that of King's College, where memories of the Bloomsbury era lingered on with Dadie Rylands, Peter Avery, Christopher Morris, Richard Braithwaite and others.

With six or seven sets of oral history washing through my life by the age of thirty it would have been a rather insensitive and myopic young scholar who did not realize that he was experiencing something special and perhaps something that should be recorded for posterity. I did not want to write a conventional history of academic disciplines, yet I had a sense that there were webs of ideas and connections which were carried by elderly individuals which should be recorded. They should be treated with the same respect and desire to save and document as was accorded to small oral cultures around the world whose myths, legends, rituals and folklore was carefully recorded by anthropologists.

It was perhaps a jumble of such motives, combined with friendship with the Director of the Cambridge University Audio-Visual Unit, Martin Gienke, which led to the first experiment to record academics on film. I had been asked to organize a series of seminars in history and anthropology in 1976–77 for the Economic and Social Research Council and this I knew would be a chance to gather together some of the leading thinkers in both fields. If they were coming for several days to Cambridge, might it not be an idea to film them, both as an archive for future generations and to show to students and colleagues who could not come to the meetings? So, I organized the seminars and two were filmed. Amongst those attending were many younger scholars who would become well known – Maurice Bloch, Raphael Samuel, Keith Thomas, Keith Hopkins, Peter Burke, and a number of more senior figures in their prime such as E.P. Thompson, Edmund Leach, Maurice Godelier, Jack Goody, Arnaldo Momigliano and Ernest Gellner.

I had assumed that this experiment, which is up on the web would be repeated by others as the technology became better. Yet, to my knowledge, it has hardly ever been duplicated. One of the few instances when something similar was done was again an idea I had. When making a six-part series for Channel 4 on the history of the world over the last 10,000 years, we assembled five leading historians and they talked for two days about various aspects of the past. The whole was filmed and is also up on the web.

The next step was probably influenced by this early filming attempt. Jack Goody had been involved in the filmed seminar and five years later, with the help of the Cambridge linguist Stephen Levinson, he arranged three seminars where Meyer Fortes, Audrey Richards and M.N. Srinivas would be filmed talking about their lives and work. Jack Goody then retired, but continued to take an interest in the project and interviewed several others for me, including John Barnes and Jean La Fontaine.

In 1982 a portable video camera which took low-band U-matic was available and so I founded the Rivers Video Project with several of my Ph.D. students, to experiment with the use of film in teaching and research in anthropology. Among the early projects were several biographical and archival films, including a film of Brian Hodgson and a long interview in June 1983 of my supervisor Christoph von Fürer-Haimendorf. The group also filmed my undergraduate lectures and some of the students made a film as part of their course on T.R. Malthus.

Now that I had started to film interviews, I could see the way in which my early interest in the gossip and interconnections of what was in effect a small tribe of British social anthropologists could be turned into a permanent record. The first interviews tended to be done by someone who knew the subject well – Goody in the cases above, Jean La Fontaine interviewing Lucy Mair, Caroline Humphrey interviewing Owen Lattimore. But gradually this faded away and I did nearly all the interviews and all the filming. About ten of the interviews were however done by friends and colleagues abroad and sent to me.

At first the focus was almost exclusively on anthropologists – with some sociologists who were also anthropologists such as André Béteille, John Barnes, Ronald Dore. Later it was extended to others in the general area of social sciences, more sociologists and some historians.

The filming continued to be done for the next ten or fifteen years on low-band U-matic and the tapes were stored away. It was not easy or necessary to edit them, but nor was it easy to show them, so very few were ever shown to students or colleagues. I looked at the collection as an archival collection, a time capsule of a vanishing world, which might be interesting in a hundred years' time. That was assuming that it would be possible to play them. Already twenty years after they were made the tapes of the ESCR seminars were unplayable in Cambridge and had to be saved by the National Film Council.

Then a set of interconnected technological changes occurred which affected many of my projects. The internet and high-speed broadband made it possible to send films across the world. Large external hard discs and new compression techniques made it possible to store and edit the films. The arrival of a Digital repository in Cambridge in 2004 made it possible to put up the interviews we had done until that date in June of that year, an idea first suggested by Mark Turin.

This development, which meant that the material could be appreciated by many more people, and the arrival of better cameras for the filming, meant that I began to increase the number of my interviews. By 2004 I had done about thirty interviews, by 2006 there were about 70 interviews and lectures. I described my project in relation to the Arts, Humanities and Social Sciences to some colleagues in King's College. The scientists Patrick Bateson and Herbert Huppert were excited by the idea and asked whether I could extend the filming to cover the natural and physical sciences.

I was surprised that very few people or organizations appeared to be doing the in-depth 'lives and works' kind of interviews I had undertaken for the immensely larger fields of the sciences. (Two sites which did have interviews on were that at globetrotter.berkeley and the Vega project at Sussex initiated by Harry Kroto). I was also hesitant about my ability to interview in fields about which I knew nothing. But they urged me to try and introduced me to a number of distinguished scientists. Shortly afterwards Gabriel Horn became an active advisor on the project as well.

I started gingerly with the scientists, filming Dan McKenzie with whom I shared a room in King's, the co-discoverer of tectonic plates and continental drift. It went well and I greatly enjoyed it. Pat

Bateson then interviewed Gabriel Horn and I filmed another session with them talking about their partnership, and I interviewed Pat Bateson himself. So, I was now sufficiently confident to interview other scientists. Having broadened it beyond arts and social sciences, there was all the world which lay between, such as economics, psychology, philosophy and so on. There were also other interesting people I wanted to film – the Vice Chancellor Alison Richard, the ex-Cabinet Secretary Richard Wilson, the musician John Rutter, the sculptor Anthony Gormley.

In fact, the series broadened out not only into politicians and civil servants, but into many other fields. There are interviews on the performing and visual arts, crafts and skills (pottery, gardening, sculpture), law, journalism, literary criticism, business, exploration, publishing, the military, filmmaking, classics, archaeology, architecture and linguistics. Whenever I heard of or met someone who seemed to have made a creative contribution to the world, whether they were a road sweeper or a College butler, up to some of the most distinguished thinkers in the world, I would try to interview them. It is a wide spread.

So, from May 2007 when the first scientists were filmed, until September 2009 when I formally retired as Professor, I filmed something like another 80 interviews and the total now stands at about 200 films.

As far as I know this archive is unique and as a number of the major figures are dead and others are getting very old, so the interviewing cannot now be done. There are some small projects, sometimes influenced by our experiments, in Germany, the London School of Economics and elsewhere. Yet on the whole there is nothing quite like this.

***

Naturally the way in which I have done the interviews reflects both my own interests and my experience of what works. My central aim has been to let the speakers tell their own story, present themselves as they wish to do, without threatening or probing or adversarial questioning. Yet in doing so, and with a roughly chronological approach, I have been particularly interested in the question of what accidents, historical contingencies, or personal traits have influenced the subjects and sent them towards a particular career. In particular, what they do has contributed to their more creative and sometimes inspired work and innovation in their various fields.

From the very start of my interest in the oral history of my seniors and contemporaries, I have been keen to learn what I should do to make me as good an anthropologist and historian as possible. What are the craft skills and what are the ways of thinking and ordering one's life which will maximize the chance of making a serious contribution to our knowledge of the world? This early interest was heightened when I found myself in the incredibly privileged surroundings of Cambridge for forty years, surrounded by the ghosts of Newton, Darwin, Rutherford and many others.

The following describes the topics I have tried to cover in my interviews. On the surface, the interviews are almost unstructured and I avoid referring to a written questionnaire as this can distract from the spontaneity of the occasion. I encourage the interviewee to talk about whatever they would like. My role is similar to a psychiatrist's, that is to say to let the subject narrate their life, in particular in relation to the obstacles and encouragements to creativity and discovery. We tend to cover the following:

when and where born
ancestry – going back as far as they like, including occupation and temperament and possible effects of grand-parents, parents and siblings
first memories and hobbies as a child
first and subsequent schools, with important teachers, hobbies, subjects which gripped them, sports and games, music, special books
university and those who taught and studied with them and interests there
first research, supervisors, mentors, influences

jobs and career and travels through life, work abroad
colleagues, friends and network of workers, partners and children
methods of working and thinking
major achievements and problem-solving during life, how they occurred, including especially important bursts of activity
administrative tasks
teaching and supervising of students
effects of their work environment (laboratories, departments, Colleges etc)
philosophy and religion
political views and activities
advice for a young person starting out in their field
specifically ask if there is anything which they would like to have talked about and I have omitted to ask about

Yet if the subject does not want to follow this order, or to answer all of these, or to add further subjects, that is fine. What I want the viewer to see is the inside of a life, told in a conversational and personal way.

***

The interviews are an intimate probing of personal experience, usually by a complete stranger who is holding a potentially threatening video camera. The subjects know that this may be seen by almost anyone in the world – friends, students, competitors, and enemies, now and in the future. This could be intimidating, especially to older subjects and for those who share a widespread reserve and distaste for talking about themselves.

I have therefore developed a number of techniques for putting the subjects at their ease. These have contributed, I believe, to the rather startlingly honest and trusting conversations which I have managed to have with a wide range of near strangers. It is worth briefly summarizing these since they could be helpful for others who help to extend this project.

1. It is important to have a fairly small and unostentatious camera that does not dominate or frighten the subject. The less intrusive the microphone the better – which is one reason why I have given up using lapel microphones. I place the camera on my knees and do not use a tripod, which can again be intimidating.
2. The room in which the interview is done is important. I avoid formal settings if possible – lecture theatres, 'offices', and seminar rooms. A room with gentle furnishings, an easy chair for both interviewee and interviewer, books and pictures and objects in the background, a pleasant view all helps. And of course, absolute silence and absence of telephones, mobiles, computers and interruptions is essential. I do not sit too close, or too far away. I sit at the same level, as I would do in any normal relaxed conversation between friends.
3. I try to develop the sheep dog technique. When gently moving a flock of sheep to its destination, a good sheep dog is mostly silent and still. Each time the sheep move in a satisfactory direction, the dog creeps forward. And then sinks onto the grass and waits attentively. It does not bark, just guides. So, if possible, I try to help the interviewee along, but only interrupt when they need encouragement or direction. I never shut them off (though I occasionally warn them if the conversation is getting into the realms of damaging speech and check that they are aware of this), but try to bring them to subjects as they are needed.
4. I always try to show interest, however little I know, or even care about the subject being discussed. What is being said is often important to the subject and has a depth which I, or others, may only realize later. They deserve serious attention and respect for what is often a summary of a life. Of course, I may verbally disagree a little, or query things, but I try always to do so in the pursuit of a common goal of understanding. Curiosity is the most important attribute.
5. It is important for there to be no sense of rush. If I want an hour of film, I allow 90 minutes, which gives time for general conversation, a cup of tea etc.

6. I used to prepare carefully for the interviews. With people in my own subjects, this was possible. With scientists, beyond reading a brief life in an encyclopaedia, I cannot really prepare. It seems to work as well without preparation.
7. I used to think that it would be good if the 'subjects' prepared themselves in some detail, and when they asked me, I would advise this. In fact, I have found that spontaneity, even if it leads to some confusion, forgetting of names etc., is better and I advise people not to think about the interview – just that it will be chronological and they can say what they like (though they can look at one or two of the earlier interviews on the web if they would like to do so).
8. The fact that there is no commercial side to the endeavour has an effect. That I am doing it without specific pay for the job and not as part of a well-funded project, is usually obvious and helps. That all the materials are freely available on the web, can be downloaded for free anywhere in the world and used in teaching and research, all adds to the trust and spirit of altruistic collaboration.
9. The absence of any bureaucracy is important. We enter into an implicit contract. I have no paper for them to sign, assigning copyright, intellectual property rights etc. It is all agreed verbally and informally in the act itself. And hence the bond of friendship is not broken.

One of the things that has developed over the years and has greatly increased the interest and usability of the interviews is the possibility of putting up a summary with some time codes to help viewers navigate to an area that particularly interests them. The summaries are often very detailed and the development of the web has again made them more interesting and reliable since one can check names, theories, and connections. It is an art form in itself, combining considerable synthetic skills, a jigsaw ability and great concentration. It is not easy, but the website gives many examples of highly professional examples which have won high praise from the interview subjects who are often amazed at how accurate and complete they are. The obvious comprehension shown in the summary further adds to the sense of trust.

10. Before the interview it is important to explain that anything that is said can be retracted or glossed later. People should not censor themselves too much. Candour and a relaxed flow of ideas are important and trying to avoid things detracts from this. I explain that while filming – before or after saying something – the interviewee can easily say 'this is not for public dissemination', 'this is confidential' or whatever. Any such passage is then excised from the version that becomes publicly available – but the original tapes are kept for posterity. I also explain that we will send them the full summary which needs to be checked for accuracy (especially names and technical terms), interpretations of statements, and also gives the person a chance to withdraw any section or passage if they wish. They may, as has sometimes happened, feel that they want to add something – some more autobiography, a clearer exposition of something technical. It is not difficult to put this into the summary either in square brackets or as an appendix.
11. The duration of time people can concentrate varies. Most people can manage an hour, and then, with a break, another hour. When the tape ends, I allow a few minutes for revival – but it is important not to lose the momentum. Some people prefer to do an hour, go away and come back some days later. This is alright, but can lead to repetition. But for older subjects (and many of mine are in their later eighties and older) it may be necessary. The older subjects also often feel more comfortable in their own homes amongst their books and belongings. This often gives an added dimension to the interviews.

If the subjects do reveal things which should be kept private for some years or even decades, these are edited out of the versions that go on the web, but the original tapes and full interviews will be preserved for posterity. Meanwhile the edited versions are put up on DSpace at Cambridge and on other websites, including YouTube, and now increasingly on websites all over the world – Scandinavia,

America, China and elsewhere. The value of the interviews is enormously enhanced by the excellent and thorough summaries which Sarah Harrison has made of them all, taking much more time and skill to do than the interviews themselves.

What has been created, through a set of accidents, is a resource for the study of a number of academic disciplines, from anthropology to molecular biology, from history to astronomy, from sociology to mathematics. It also provides rich material for the study of British academic life and institutions in the twentieth century. Furthermore, for those interested in the conditions of creativity and discovery, it is a unique archive.

The project has developed as a result of a set of accidents and through the help and support of many people, in particular Sarah Harrison, Mark Turin, Jack Goody, Gerry Martin and many others. This project to publish the full interviews of selected individuals was the idea of Esha Béteille, whose support and advice have been indispensable. I would particularly like to thank Radha Béteille who it has been a great pleasure to work with. Her transcripts and editing have been extraordinarily well done and I am deeply impressed and grateful for her contribution. Institutions such as Cambridge University (DSpace and the Streaming Media Service in particular) and King's College, as well as the British Academy, the Leverhulme Trust, the Firebird Foundation and others have also made all this possible. It is hoped that others will take the task on through the 21st century.

PART one

Andrew Fielding Huxley. Photograph courtesy Alan Macfarlane.

Andrew Fielding Huxley

Andrew Fielding Huxley, OM, PRS (1917–2012) was an English physiologist and biophysicist. He was born into the prominent Huxley family. After graduating from Westminster School in Central London, from where he won a scholarship to Trinity College, Cambridge, he joined Alan Lloyd Hodgkin to study nerve impulses. Their eventual discovery of the basis for propagation of nerve impulses (called an action potential) earned them the Nobel Prize in Physiology or Medicine in 1963. They made their discovery from the giant axon of the Atlantic squid. Soon after the outbreak of the Second World War, Huxley was recruited by the British Anti-Aircraft Command and later transferred to the Admiralty. After the war he resumed research at The University of Cambridge, where he developed interference microscopy that would be suitable for studying muscle fibres.

In 1952, he was joined by a German physiologist Rolf Niedergerke. Together they discovered in 1954 the mechanism of muscle contraction, popularly called the 'sliding filament theory', which is the foundation of our modern understanding of muscle mechanics. In 1960 he became head of the Department of Physiology at University College London. He was elected a Fellow of the Royal Society in 1955, and President in 1980. The Royal Society awarded him the Copley Medal in 1973 for his collective contributions to the understanding of nerve impulses and muscle contraction. He was conferred a Knight Bachelor by Queen Elizabeth II in 1974, and was appointed to the Order of Merit in 1983. He was a fellow of Trinity College, Cambridge, until his death.[1]

I

Alan Macfarlane (AM): It is a great pleasure and privilege to be able to talk to Sir Andrew Huxley. May I call you Andrew? Is that all right?
Andrew Huxley (AH): Yes, by all means.

AM: Andrew, when were you born and where?
AH: I was born on 22 November 1917, in Hampstead.

AM: I always ask about parents or sometimes even grandparents. In your case I would like to start with your grandparents because, on one side at least which I know about, they were extremely distinguished. Can you tell me something about your grandparents, the Huxleys?
AH: Well, T.H. Huxley[2] was my grandfather. I was disproportionately young to be only a grandson because my father was married twice and I was the youngest son of the second marriage.

Most people knew their grandfathers, but I did not know either of my grandfathers. T.H. Huxley died in 1895 more than 20 years before I was born. My mother's father was a Bruce. He worked in the city in Social Affairs on the London County Council and then retired. He died in 1908, so I never met of them.

AM: Then in a slightly more distant way I noticed on your family tree that you were related to Matthew Arnold?[3]
AH: Yes, well that is a connection rather than a relationship. My father's first wife was the niece of Matthew Arnold. She was a granddaughter of the Thomas Arnold[4] who started Rugby School and set the fashion for other schools.

AM: Are you an admirer of Arnold's poetry?
AH: A little bit. I do not read much poetry but I have enjoyed some of it. He published a collection of English poetry which I dip into.

This interview was conducted by Alan Macfarlane on 5 October and 7 November 2007. The transcription for this interview has been provided by Radha Béteille, for this volume.
These interviews are available for viewing on:
https://www.sms.cam.ac.uk/media/1123362
https://www.youtube.com/watch?v=WdL-81i3Qg4&t
https://www.youtube.com/watch?v=qL3aTfljBXE

AM: And then moving down to your parents' generation. Can you tell me something about your parents?
AH: My father was Leonard Huxley. He was a literary man but quite knowledgeable about science. He died in 1933 when I was only 15.

AM: What sort of character was he? Do you remember?
AH: He was very friendly and gentle.

AM: Encouraging to you?
AH: Yes.

AM: What sort of person was your mother?
AH: She was extremely good with her hands and eyes which I have inherited to some extent. She was a widow for a long time. She died in 1994 at the age of 104 having been a widow since 1933 now. She was physically very active my father wrote a short poem to her and addressed her as an Oread nymph,[5] as I say a mountain nymph. We took mystical holidays on the west of Scotland.

My mother's father and my mother's mother's father had taken shootings on the west of Scotland from the 1880s onwards. Then in 1902 they bought a holiday house at Connell Fairy. It is just a few miles north of Oban in a very attractive country.

When I was small, we regularly went there for summer holidays, which of course were plagued by midges and then later we took to taking our holidays in April. No midges! Sometimes very cold – usually snow on the hills, but we always enjoyed it. Nowadays I only go on holidays when one of my children takes me on holidays and very often that's on the west of Scotland. This April one of my daughter's took a self-catering house near Noch Torrin, a good bit further north in the open and we had a lovely time. I still walk a fair amount but I cannot climb mountains in the way that I used to or it was hill-walking. I never went in for rock climbing.

AM: Tell me tell me something also about your mother, she was your father's second wife. As you mentioned briefly earlier, by his first wife he had several sons at least two of them are internationally known. Can you tell me something about Aldous[6] and Julian?[7]
AH: Julian was 30 years older than me and Aldous several years younger. They were more like uncles than brothers. We saw them

quite often before the war. We were living in Hampstead. Aldus, of course travelled a lot, but when he was in London, he came to visit us quite often. Both of them were very good company in different ways. Julian was full of stories very often laced with impropriety. Aldous was always interesting and friendly. He had very poor eyesight. He had for a time been actually blind and had learned Braille. The one great advantage of reading Braille was that you could read in bed without your hands getting cold!

AM: He wrote a book about it didn't he? Do you think either of them influenced your future career in any way – the fact that Julian for instance is a very great biologist and zoologist?
AH: I do not think so. My sort of science has been very different from Julian's. Julian was an essentially an observer primarily of birds and interested in their behaviour. He founded the science of animal behaviour. But I have really had very little interest in the behaviour of living animals.

T.H. Huxley wrote a short autobiography[8] in which he says his boyhood ambition was to become an engineer, but he went into medicine because he had a brother-in-law in medicine. He says the only part of the course that really interested him was physiology because it is mechanical engineering of living things and that's an exact description of my interest in physiology which has been my main work. But I think engineering is in the family.

When I was a boy, my interests were all mechanical – playing with mechanical toys – I had quite a lot of Mechano. It was sold in sets but my mother did not buy the sets for us. She bought the separate parts so that we had to design our own things with them – railway sets, gauge zero, gauge one-and-a-quarter inches, which we put out once a year in the winter holidays. We took over the spare room at our home with a quite elaborate layout.

My two-year-old brother David,[9] was more interested in organization and layout. I made a number of mechanical things related to the railways. When I was about 14, my parents gave my brother and I, a metal turnings crew cutting lathe, which I still have. It is in the garage. I made a lot of things on it. I had it in my lab when I was doing experimental work and made quite a lot of my own equipment. I made quite a number of improvements to the lathe such as putting scales on its movements and fitting an arrangement for automatic feed in the cross direction as well as the forward direction and virtually, like all lathes, it had the clutch for the lead screw. The screw that moves the tool when you were cutting the thread, but the clutch for it, was on the on the same shaft as the lead screw. That meant that if you cut a thread you would have to take several cuts to get it deep enough and when you went back to take another cut, it would not fit into the existing groove, unless the thread that you were cutting had a pitch that was some exact fraction of the pitch of the lead screw ten to the inch. I, therefore, made another clutch on the back end of the shaft that carries the work that you are turning so that there when it was declutched and re-clutched again after moving to the back to take a deeper cut it automatically engage, after an exact number of turns. So, you deepen the cut whatever the pitch.

AM: Was this patented?
AH: No, I think it was so obvious that I could not patent it.

AM: It sounds like a very good idea. You mentioned this was at about the age of fourteen that you were given this and if we could move to your schooling. Do you remember your first school?
AH: Yes. It was the junior branch of University College School which was in Holly Hill at the top of Hampstead Heath, about half a mile from our home. I have walked there every morning and then I was at the senior branch for a year. The junior school was rebuilt while I was at it and while it was being rebuilt, we moved into the gym of the Upper School in Frognal. I still keep in touch with them. They send me the annual report and they had a dinner celebration. They are celebrating the hundredth anniversary of their move to Hampstead.

They had originally been in Gower Street next door to University College itself and my father was there as a schoolboy and so was my mother's father. We have a long connection with the school. But after one year, my mother got me moved to Westminster School where I was for the rest of my schooling. I had been doing classics for the first two years of Westminster. My mother, this was after my father had died – knew my serious interests were in science and against opposition from the headmaster got me transferred to the science side and used the phrase that I was forsaking virtue for pleasure!

I was extremely well taught in physics by J.S. Rudwick, whose son was Martin Rudwick,[10] the well-known historian of science. I got in touch with him again when he was an undergraduate at Trinity. He was then a junior research fellow at Trinity for a time and now comes to Trinity fairly often so I keep in touch with him.

AM: Whether any other teachers who influenced you a lot at Westminster?

AH: There was a Mr Claridge who taught French. But I had been taught French by Mademoiselle Francis who gave classes at the home of Sir Thomas Lewis,[11] the famous physician. This was not too far from my home where I was brought up. I first went to her classes at the age of six and then my mother arranged for her to come for an hour a week teaching French to my brother and myself. She taught me not only French grammar but French pronunciation. She told me what to do with my tongue in order to make French sounds instead of English sounds. All our vowels are really diphthongs. As a result, I have a fairly good French accent and in Paris once I was mistaken for a Frenchman from some other part of France. But it has helped me in other languages as well.

AM: What other languages are you fluent in?

AH: I learnt German. My brother and I had a spring holiday in Salzburg. I did read quite a lot of history of science in German and the other language that I did learn at one time to a rather moderate level was Japanese.

I was invited to a physiological Congress in 1965 in Tokyo and before that I met a distinguished Japanese biochemist who visited England in 1960 and came to visit Alan Hodgkin,[12] my mentor and myself at that time and then he introduced to me one of his students who did a PhD under me, Makoto Endo.[13] We have kept up ever since. My wife that I went to this Congress in Tokyo in 1965 and this Makoto Endo looked after us and took us to all other places in Japan such as Kyoto and we visited a lot of temples.

AM: Did you like Kyoto?
AH: Yes, it is very impressive and beautiful.

AM: Did you go to Nara which is just next to Kyoto?
AH: On a later occasion actually. I have been to Japan a number of times since, almost always based on scientific meetings. But I did learn some Japanese and I actually gave a lecture in Japanese.

AM: Good heavens!
AH: At that time, I had not started to learn the real Japanese script which was composed of Chinese characters borrowed from different dialects of Chinese at different times. Each character is associated with several different pronunciations and most words consist of two characters. The character is pronounced quite differently according to which Chinese dialect it was borrowed from in this context.

AM: Did you actually learn to write some Japanese as well because there is *hiragana*[14] and *katakana*[15] as well?
AH: Yes, after coming back I did settle down. I learnt fair amount of the two syllabaries. *hiragana* and *katakana* – not alphabets, but syllabaries because they have about 50 characters and each being a consonant plus a vowel. There are no separate categories for 'L's and 'R's and of course Japanese are notorious for mixing them up. I did see in print the word 'cerebration' instead of 'celebration'! I know after coming back from Japan I went to a number of lessons at School of Oriental and African Studies [SOAS] and got a book to learn Japanese from one of my daughters. She had the book and tested me on it but I really virtually entirely forgotten it now.

AM: You did it just out of interest? There was no practical reason for learning Japanese?
AH: It was interest. As a language, it has some resemblances to English but it is very different and of course the writing system is entirely different. Each character is made up of component elements but you recognize the character as a whole in the same way that you recognize an English word as a whole. I remember reading an article in *Nature*, a number of years ago, claiming that we did not read words as a whole, but we read them letter by letter and if were

reducing the intensity of the light with which you read until you could hardly read, then you did have to read words letter by letter. The fact that we do read a whole word at a time struck me again. I used to read aloud to my wife in bed before going to sleep and at one time I read some of Steven Runciman's[16] book on the crusades and of course it has a large number of place names in it which were quite unfamiliar to me and they always brought me up standing. I could not read them freely. I had to read them to work them out letter by letter and then pronounce them.

AM: That's good evidence. This came out of your interest in language like the French at school.

Whether were there other things at school? At one point you mentioned Salzburg so that immediately makes one think of music. Were you interested in music at any time?

AH: Yes, and I did learn to play the piano at a rather low level but it was always at a low level. When I was an undergraduate, I played quite a lot of chamber music with friends who played the violin, viola cello and the clarinet.

AM: Your music, your clarinet playing was good enough to play with a small group.

AH: Yes, easy classics was my level.

AM: Now are there any classics which you are fonder of than others, most fond of? Do you have any favourite composers?

AH: I think Bach[17] was my favourite. Schubert[18] comes high. My eldest daughter who shares this house with me has quite a good singing voice. I used to accompany her singing good Schubert songs and one or two Bach songs and I was fond of Brahms.[19] Schubert's late String Quintet is one of my favourites and one or two Beethoven's[20] Lake Quartets.

AM: All German, by chance? That is not surprising since most great composers are German.

Towards the end of Westminster, you are now fully specializing in science. Was it, particularly, physics?

AH: Yes.

AM: Was it particularly physics?
AH: Yes. I had a little biology. I think I did dissect a frog and a rabbit. But my interests at that time were really entirely on the mechanical side. I came up to Cambridge aiming to specialize in physics and become an engineer. But in the Cambridge system I had to take for my first two years three different sciences. Naturally, I took physics and chemistry and when I was wondering what to do for my third science, I went to stay with a friend, a few years older than me, who lived not far from us in Hampstead whom I got to know as boy. This was Benedict Delisle Burns.[21] Later, he became a Fellow of the Royal Society. He is best known for recording from cells in the cerebral cortex. He was then doing Part 2 physiology and he recommended physiology as my third subject on the grounds that even in the first year I would be learning things that were still controversial unlike physics and chemistry all cut and dried. I did pick on it. I think because I had come up with a major scholarship in physics my director of studies thought I did not need supervisions in physics. I mostly had supervisions from William Rushton.[22] A very interesting person. He took me on solo supervisions which is a bit unusual in Cambridge since you usually have two or three together and we got on very well.

AM: He was a good teacher?
AH: He was a very good teacher and very amenable. I decided to specialize in physiology for my Part 2 and the First Lord Adrian[23] – E.D. Adrian who became head of the physiology Department in 1937 when I was just finishing part one with the Tripos and he told me that if I wanted to make a career in physiology, I ought to become a medically qualified because in those days all teaching posts in physiology went to people with medical qualifications. This is no longer true. I did register as a medical student and I have done the necessary physiology to get exemption from the qualifying exam in physiology in the Part 1 of the Tripos, but I had not done any anatomy. So, I spent my third year doing anatomy lectures and five parts of the human body that have to be dissected – the head and neck, thorax, abdomen, arm and leg. Of course, there are only three terms in the year. I did one part in each term and one part in the

April vacation and one part in the summer vacation. We always did the dissections in groups of two or three. In the April vacation when I went to medical school at Barts to dissect the thorax – somebody whom I had known slightly at Trinity, a contemporary of mine, Andrew Barlow – do you know any of the Barlows?

AM: Yes, I have heard the name. I think one of them was at King's I have a feeling but anyway.

AH: Horace Barlow[24] was at King's. He would have been a Fellow of Trinity and I would see him regularly. But this was Horace's next elder brother and we got on very well. Neither of us was very ready with conversation. He and I found ourselves during part two physiology in 1938–39 – very small, very agreeable class with only 12 of us. But in our third year we were both as I say during anatomy and we did thorax trunk clear parts.

AM: Who were your teachers? I mean Lord Adrian you directly mentioned. Was he directly teaching you?

AH: He lectured to us in in our part two class. Before that when he was in Cambridge, he had a Royal Society professorship and not a teaching post. I never had supervisions from him. I mostly had supervisions from William Rushton. But in my final year when doing Part 2, I did have supervisions once with Alan Hodgkin and that was how I first got to know him.

AM: How much older than you was he?

AH: Four years.

AM: It was rather strange being say 20 and he was 24. Something like that?

AH: Yes.

AM: Did you take to him straight away because you had a long association?

AH: Yes, because I knew him very well and that summer of 1939, he made arrangements to go to the Marine Biological Laboratory of Plymouth, to do experiments on this very large nerve fibre that squids have. They have a pair of these fibers and the reason they are large, speaking evolutionarily, is that the crude way of getting a high velocity of conduction, is to have a large diameter. The speed of conduction goes up with a square root of diameter and squids need to activate all the muscle of their mantle with which they make this squirt of water with which they propel themselves backwards. They need to activate all their muscles synchronously so they have this artificially high conduction speed in their nerve fibers in order to do it. They have this pair of very large nerve fibres which are very convenient for experimental purposes. They are about half a millimetre across while the largest fibres in us are only about a 50th of a millimetre across and they get a relatively higher velocity of conduction by a different trick. The velocity is limited largely by the high electrical capacity of the surface membrane due to it being very thin. It is just a bi-molecular layer[25] and all vertebrates for the fast-conducting fibres have extra layers of membrane on the outside so that the membrane is enormously thicker than the actual excitable membrane and they have to have little gaps in this thick membrane in order to have the actual excitability which could not work through all these membranes in series.

This sort of nerve fibre had been investigated by a Japanese cutter in the 1920s. He dissected out single fibres. The general feeling in the West was that nobody except the Japanese were skilled enough with their hands to dissect single fibres. An American couple, Ramsay and Street, who were husband and wife, and I think Sybil Street did teach herself to dissect single fibres and well I think I said I inherited my mother's ability with her hands. Then another Swiss, Alex Von Muralt[26] had taught himself to intercept single fibres and his pupil Robert Stämpfli[27] came to work with me. The contact was originally because Alan Hodgkin had known von Muralt and from them I learnt the technique of dissecting single nerve fibres and later on I learned to dissect single muscle fibres that were a good deal more difficult from muscles of frogs. All my experimental work has been on single fibres from frogs, and muscle fibres, partly from vertebrates but also from crabs and lobsters, because their muscle fibres are very much larger in diameter and all our fast muscles crossed by what are called striations. In vertebrates these striations are separated by only a few wavelengths of light so it is difficult to get a good image of them.

But crabs, lobsters and insects have fibres with very much wider spacing which can be studied satisfactorily with the ordinary light microscope. There is virtually no absorption of light by the striations and the ordinary microscope is designed to give a good image for regions that absorb light, but the ordinary light microscope can be used to look at regions of high refractive index, because they act like little lenses to concentrate the light above them. If you go out of focus upwards, bright regions are regions of high refractive index and if you focus low, dark regions are regions of high refractive index. It became customary in the 19th century to use the dark focus position – what are called the A bands[28] of muscle – 'a' for anisotropic,[29] because they affect polarized light in a particular way but they have a higher refractive index than average so they are dark when focused down. They got commonly called the dark bands, though they are not intrinsically dark.

I did a certain amount of work on these invertebrate muscle fibres but they are not fully excitable in the way that a vertebrate muscle fibre is. A vertebrate muscle fibre is activated by its nerve at a particular region and the activity spreads along the muscle fibre in very much the same way that it spreads along a nerve fibre. Trying to remember what sequence things happened then....

AM: We got you. Alan Hodgkin going down to look at John's squids in 1939. You know presumably at the end of more or less of your undergraduate degree the war came.
AH: Yes.

AM: And then what happened in the war?
AH: Well first just what happened at Plymouth. Actually, Alan Hodgkin already had hints from recordings on what he had learned to isolate single nerve fibres from crabs. They have largish nerve fibres, not as large as the squid ones, but the same type as squid ones. They are something like a fifth of a millimetre across unlike the half millimetre of the squid fibres, but they have only this thin membrane on the surface so they could not conduct with a moderately high speed. He made a number of very important observations on these fibres. In 1937–38 he was in the United States and met Kacy Cole[30]

who was already doing experiments on the giant nerve fibre which had been discovered by J.Z. Young,[31] a professor from Oxford. He became professor of anatomy at University College London where I saw quite a lot of him.

AM: Didn't he come to Cambridge later?
AH: Young? No. Not this Young. There are a lot of Youngs.

AM: No? Sorry. So, he was at London?
AH: But anyhow in summer of 1939, Hodgkin went to J.Z. Young who had discovered this giant nerve fibre in squids but had never done anything really significant experimentally. He measured its speed of conduction which was high compared with small diameter fibres but Hodgkin went to Plymouth to study them and invited me to join him in August. Earlier, I had been on holiday in Scotland and so I did join him at the very end of July and between us, we managed to get these fibres set up so that we would put an electrode down inside them. We had them hanging vertically in a trough of seawater which has the same composition as the body fluids and we would stimulate it at one end and record the impulse as it went down. In order to get a good recording, we needed keep down the resistance of the inside of the tube that we put inside. It was a glass tube drawn out with an outside diameter of a tenth of a millimetre that was a lot smaller inside. We had to have it filled with the fluid, I think we use seawater again at the same concentration as the squid was accustomed to but that gave it a rather high resistance and that together with the input capacity of the amplifier would have distorted the action potential.

I threaded a bit of silver wire down the inside the glass tube electrode to keep down the resistance and then a trick that Alan knew was to prevent the silver from becoming polarized by passing current. He passed a current making the silver the positive electrode which dissolved some of the silver and made a precipitate of silver chloride – very insoluble on the outside of the silver, so that the silver metal was in contact with a saturated solution of one of its salts and that made it a non-polarizable electrode. It could pass small but appreciable amounts of current without the voltage being distorted by polarization. That enabled us to record the impulse from the inside of the squid fiber without distortion and when we put it down first, we recorded the resting potential. That was already well-known since the 19th century that the insides of nerves and muscle fibres are electrically negative to the outside, the actual value of the negativity was not known but we found that it was in these damaged fibres about 50 millivolts negative inside. When we stimulated the fibre at one end and the action potential went past the tip of this electrode it did not merely come up toward the potential of the external fluid which was what all the textbooks told us at the time but it went very substantially positive. We published this in a short note in *Nature* describing the observation, but with no discussion and no conclusions.[32]

AM: When was this?

AH: Well, this was right at the beginning of the war, it was in October 1939. We had no explanation for the overshoot inside going positive during the impulse. During the war Alan Hodgkin and I met and speculated about what the mechanism of this was and we published a paper, a full-length paper in the *Journal of Physiology* in 1945 in which we put forward four possible explanations. All of which turned out to be wrong and with hindsight I was reminded of a thing with my grandfather T.H. Huxley said when he first made himself master of the central idea of the *Origin of Species*, his response was I was stupid not to have thought of that!

The explanation for this overshoot is in fact very simple. It was known from beautiful experiments by Kacy Cole before the war measuring the impedance of the membrane at a range of high frequencies. It was known that the membrane had this very high capacitance and the resistance of the membrane fell to a very low value when an impulse went along. It would become permeable to ions and with hindsight it was very stupid of us not to realize that this high permeability is, in fact, specific for sodium ions. Sodium ions are much more concentrated outside the fibre than inside so they diffuse in during the action potential carrying their positive charge and making it go positive.

AM: Meanwhile the war is going on. We will come back to this and your later work with Hodgkin in a minute. But what happened when the war started? You would have just finished your degree? Did you stay?

AH: I had registered as a medical student and I had done the necessary qualifying exam. The obvious thing was to go ahead toward getting qualified and now at that time the Regius Professor of Physic at Cambridge,[33], physic not physics, was John Ryle,[34] a very distinguished physician and he had been at Guy's Hospital[35] in London. He loved teaching but he had not got anyone to teach and there were two or three of us who needed teaching. He ran an introductory clinical course at the Addenbrooke's Hospital[36] in Cambridge for the first three months of the war and I was on this team.

I found clinical work very interesting. After that I moved to University College London and went on with clinical training until teaching stopped because of the bombing of London in June 1940. Meanwhile, in Cambridge, through John Ryle, I was introduced to Robert McCance[37] and Elsie Widdowson.[38] I do not know if you know their names, they were the leading nutritionists at that time.

They had got together at King's College London before the war and had produced the first table of calorie values of foods and so on. The natural thing for them to do at the beginning of the war was to conduct an experiment in rationing. Through John Ryle and being a medical student, I met them and they recruited me as one of their guinea pigs. For the first three months of the war and in fact a bit longer, I was on this experimental rationed diet. It was like our wartime rations except that the rations that we had were of leek, butter, sugar, milk. About half of what we had during the war. Bread and potatoes were not rationed. It was not until after the end of the war that bread and potatoes were rationed. It was very difficult to organize because different people's requirements of total calories varied so enormously.

Nowadays our calorie intake is less than half what it was at that time but we were weighing everything we ate and we were weighing ourselves some every day and we were weighing the products of our bowels and having them analysed at regular intervals. At that time, I was taking a lot of vigorous exercise. In New Year 1940, the first winter of the war, McCance took Elsie Widdowson and McCance's own son, then aged about 14 and myself and I think one of the other guinea pigs to the Lake District to take the girls' exercise. McCance was very energetic and so was I in those days and we had one marvellous walk. We were staying in Langdale at a house I had stayed in before because it belonged to George Trevelyan[39] whose wife was a step cousin of mine. I have known the Trevelyans through that connection, it was very important to me. I loved this walk, set off 6.00 am in the dark. A walk of, I think, 37 miles, 7,000 feet up and down got back up at 6.00 pm in the evening, there was snow on the ground but not a cloud in the sky the whole day – best walk I have ever had.

AM: Wonderful and then you will no doubt be weighed and measured when you got back?
AH: Yes, we were still on these rations and we were chewing brown bread all the way!

AM: What was the purpose of this holiday – to actually do a scientific experiment or just for fun?
AH: It was checking that we were still in good shape. It was not clearly a scientific experiment, but it was it was an experiment.

AM: I don't know how old G.M. Trevelyan was. Did you know him at all?
AH: Yes, I saw a lot of him both as an undergraduate and later. He asked me out to lunch many times so I would say the connection was that his wife was a step first cousin. He invited me regularly to his holiday home in Northumberland and through that I got to know his brothers – Charles Trevelyan[40] who occupied the family house at Wallington. Through that connection what we went shooting on the Wallington grounds, I got to know the Wallington family and the youngest son Geoffrey, is one of my closest friends now.

II

AH: Bernard Katz, with whom Alan Hodgkin and I published later, was a very distinguished biophysicist worked, came to join A.V. Hill at University College in about 1930, escaping from the Nazi persecution of Jews. He had then gone to Australia just before the war on a research fellowship. The future wife of a colleague of Katz's was at a lecture that Katz gave in 1940 in which he said that the overshoot might have been due to the membrane becoming highly permeable to sodium ions. But when she spoke to him after the lecture, he had quite forgotten that he said it. I had a similar experience of having given a plausible explanation to someone at a party of non-physiologists and was talking to some of them about my work, after that conversation, I had a strong memory that I had given a plausible explanation for the overshoot, but I could not remember what it was. The only plausible explanation is the membrane becoming permeable to sodium ions. It looked as though Katz and I both had the right idea but had somehow repressed it.

The obvious thing after the war was to measure the electrical change in solutions with different amounts of sodium ions. In 1946 the Marine Laboratory at Plymouth was out of action because it had been bombed during the war. But in 1947 Hodgkin and Katz went there and did some preliminary experiments which continued into 1948. I would have joined them but I was married that summer and was on my honeymoon. But they showed very convincingly, without a shadow of doubt, that the peak of the electric change during an impulse in the membrane became highly permeable to sodium and that idea became universally accepted and it now seems terribly obvious.

AM: Who did you marry?

AH: Two of my undergraduate friends were members of the Barlow family, direct descendants of Charles Darwin. They had a big house at Wendover and had a dance every year. I went to one of these dances just before the war and then I went again after the war. It was one of these dances at New Year 1946 that I met my future wife. She was then a second-year undergraduate at Newnham and we became engaged that summer. We put off marrying until she got her degree, so we married in the summer of 1948. She died in 2003. We had a very happy married life.

AM: Did she support you in your work?
AH: She was never directly involved in my work although she had taken physiology in Part One of the natural science Tripos, but she had many other interests. When we were engaged, she said she wanted to have six children as she was one of six. I did not take her seriously, but we did have six children and they occupied a substantial part of her life! She was also involved in a lot of village activities and charities, also was a magistrate. We had separate activities.

AM: When you became Master of Trinity, presumably she was having to entertain?
AH: She was a good social entertainer and extremely good at making people feel at home. It was a regular thing, the Master of Trinity had lunch parties for scholars of the college – a first year scholar coming to lunch at the Master's Lodge was a bit terrifying, but Richenda was able to put them at ease as soon as they came in.

AM: You mentioned her name
AH: Her name was Richenda. When I first heard it, I thought it was an Italian name, but it is a female equivalent of Richard. It is a common name among Quakers. Her ancestors were Quakers. Her grandfather, Edward Reynolds Pease, was a firm atheist, who left the Quaker fold. Richenda's father was also a strong atheist and refused to enter a Church.

AM: What about Richenda herself?
AH: She and I were both agnostic (a word invented by my grandfather).

AM: What a lovely thought! So, you were and still are an agnostic – in that one cannot either prove or disprove the existence of God?
AH: Yes. I am very conscious that there is no scientific explanation for the fact that we are 'conscious'. I am aware and I feel and see things.

AM: What about officiating a term, religious ceremonies at Trinity Chapel? Did you feel any difficulty about that?
AH: I found no difficulty in officiating at Trinity Chapel as Master.

AM: I have asked Sir Martin Rees and he said that he didn't feel any difficulty, he is also an agnostic.
AH: The Master has a stall at the west end of the Chapel. That was advantageous on one occasion when during a sermon, when everybody was attending the sermon and looking to the preacher. I noticed that one of the candles on the altar had split waxed down and there was a fire starting on the altar. I was able to touch the Chapel clerk on the shoulder, sitting in front of me and drew his attention to it, so it was dealt with.

AM: So, you saved Wren's Chapel.
AH: It is not Wren's Chapel. I think the Chapel was financed by Mary Tudor.

AM: Really, I have always gone around telling visitors that it has nothing to do with Wren, it is just a library.
AH: The old library is Wren and I think it is the only bit of Wren at Trinity.

AM: Though the Chapel is clearly much later than Tudor. Inside – it is a box-like shape – looks 18th century rather than Tudor.
AH: All the panelling and woodwork is relatively new, put in by Richard Bentley, who was a tyrannical Master in the 18th century who had a war with the Fellows. The stained glass on the windows is Victorian. There were Victorian paintings on the walls, but they have been covered over with a white paint that can be removed if some future generation wants to see them.

AM: As we are on Trinity, although it is out of sequence, how long were you Master?
AH: I was Master of Trinity for six years. I succeeded Alan Hodgkin, my mentor, who had also been Master for six years. The Master is usually appointed to age 70, but the Fellows can prolong up to the age of 75. If Hodgkin had been prolonged, I would already have been over 70 when the Mastership became vacant. But he

was not prolonged, I am not fully clear why. His predecessors have been prolonged since this system came into operation. When I was approaching 70, I asked to be prolonged for two years, not the maximum of five. That was granted so I was Master as Hodgkin had been, for six years. The Lodge is a very comfortable place to live in and has splendid rooms for entertaining.

AM: Did you enjoy being Master?
AH: Yes. The duties of the Master are less in Trinity than in most colleges. In most colleges, the Master is at least a member and probably a chairman of most of the important committees. But in Trinity, the Master is appointed by the Crown and the Fellows have no control over who they get. They give the Master minimal power and he chose the college counsel which meets every week in term time. When there is a meeting of all the Fellows, he normally chairs that. The duties of the Master are mostly ceremonial, a lot of entertaining and being entertained. My weight shot up! My wife and I gave up eating breakfast and in spite of that I went up from 10.5 to 12 stone!

AM: Was this the period or before the period when Trinity became exponentially wealthier as a result of various good investments?
AH: The wealth of Trinity – there was a benefactor in the early 20th century. We made a lot of money out of the development of Felixstowe as a container port. The Senior Bursar had bought quite a bit of land inland from Felixstowe in about 1930, thinking it would become valuable for housing development. There was then just one private dock in Felixstowe at that time and then Felixstowe docks were developed by the next Senior Bursar, John Bradfield, and it became immensely valuable. Another thing that made a lot of money was the Cambridge Science Park which was the first in Britain, built on land which had been college land before the foundation of Trinity. Trinity was founded by King Henry VIII in 1547. It was formed by combining two much older colleges, King's Hall, which was the second college in Cambridge following Peterhouse and Michaelhouse was the third college in Cambridge. This land to the north of Cambridge had been in a position of King's Hall which

John Bradfield developed as the Cambridge Science Park and that has been a money-spinner.

AM: Then they accidentally bought the farm where the Channel tunnel came out on this side of the town. A good set of serendipitous investments.

AH: Indeed. The College had an income that exceeded the needs of the College and a large part of it goes to providing bursaries to impecunious pupils who come up to Cambridge, to any college, not only to Trinity. We also make benefactions to other colleges to allow them to have Fellows and such like.

AM: Let us go back to your career about the time you married, what was your position?

AH: I was a Research Fellow of Trinity. I had become a Research Fellow in 1941 during the war. Trinity had in the ordinary way competitors for these research fellowships which had to be put in a full-length dissertation. Due to the war that rule was relaxed and I was awarded the fellowship on the strength of the dissertation of one and a half pages mentioning this observation that Hodgkin and I had made of the size of the nerve impulse in the fibres of squid, which we had done before the war. After the war, I was appointed, not as a lecturer in the university, but as a demonstrator. I have always had an interest in microscopes. Most of my work was demonstrating in the histology classes. In Cambridge, histology and microscopic anatomy are taught in the physiology department and not in anatomy. The two departments have now been amalgamated, but in those days, they were quite separate.

AM: You mentioned that you were a demonstrator. Presumably, did you become a lecturer after that?

AH: I became a Reader eventually. Before that, it was a mixed research and teaching post.

AM: ADR?

AH: Yes – Assistant Director of Research. Initially, I had this junior research fellowship at Trinity which would have continued after the war, then a senior research fellowship, finally a teaching post and a lectureship in the college. Then, in 1960, I was more or less offered the post of Professor of Physiology at University College London. And A.V. Hill, whom I have already mentioned, was then working in retirement in University College London heard that I was being approached. He wrote to me saying that he hoped I would accept because he had been in a similar situation in 1919 – he had come back to Cambridge after the war and was comfortably established with a university post and a college fellowship when he was offered the chair of physiology at Manchester. Lord Rutherford, the physicist, had just come back from Manchester to Cambridge and A.V. Hill asked his advice. He quoted this advice to me and had never regretted it – 'Cambridge is a splendid place when you are young and Cambridge is a splendid place when you are old, but for the middle of your life, for God's sake, get out!' I followed his example and took the chair in London. I was told that London University had a rule that its professors should live within 30 miles and I said I intended to stay living in this house which we already owned and would travel up and down twice a week. They accepted because I would have a London address which was deemed sufficient.

My system was to go to London early on Monday morning, stay Monday and Tuesday nights in London, home for Wednesday night, back to London Thursday morning and home late Friday for the weekend. When in London, I did not have a flat of my own but used bed and breakfast places close to University College. So, I stayed working until nearly midnight and was back at work before nine the following morning, and I probably spent more hours in the department than anyone else. I think it was the best way of doing a London job. Nowadays, there are huge numbers of people living in the Cambridge area and commuting daily either by train or by car. Roads are terribly crowded and so are the trains. I came back the other day on a 6.15pm train from London and there was hardly standing room on it.

AM: You mentioned your pattern of working, going in early and so on, again it is a bit out of time, but I have often asked people whether they have any advice or description of how they normally worked – their normal working day. Any advice to younger people about – because as you know, in Cambridge as elsewhere there are

constant counter-pressures towards administration and other things. It is very difficult to concentrate. So, I wondered how you managed to achieve so much in your work?
AH: After I ceased to be head of the department, the Royal Society gave me one of the research professorships. These are highly desirable posts which provided a salary with a research worker and research expenses and the cost of a technician and a secretary. I held this until I reached retiring age in 1983. I then carried on without a salary until I became Master in 1984.

AM: How old were you when you got the Royal Society professorship?
AH: I got it in 1969. I was born in 1917, so I was 52.

AM: Were you at that time still in London or back in Cambridge?
AH: I was living here in Cambridge, but I was working in London as the head of the physiology department and with the Royal Society appointment I continued my research at University College and continued travelling up and down twice a week as I have said and went on doing that until I retired in 1983. I then went on to the gap year after which I became Master of Trinity.

AM: I am not sure if this is the work on squids that you and Alan Hodgkin got your Nobel Prize for? Could you describe the events surrounding that Prize? That was 1963, wasn't it?
AH: Yes. The Nobel Prize can be given up to three persons, but not more than three. Alan Hodgkin and I shared prize with Sir John Eccles whose work, though different, it was concerned with ionic movements through nerve cell membranes – fibres coming from another cell. In retrospect, I am sorry that Cole, whom I mentioned, would have possibly been more appropriate if he had shared it with us. Our work, to some extent, did depend on his remarkable technical achievement. The events surrounding the Prize are very enjoyable. We were put up in the Grand Hotel in Stockholm and took our three elder children with us. That was a tradition that had been started by Dick Sing. I had known him as an undergraduate at Trinity and I think he had his Nobel Prize for inventing a method of separating proteins. But there are several banquets, but there was an actual Nobel banquet. I had got to know two of the senior people in

physiology in Stockholm, Granit and Samuelsson. Samuelsson had worked with Adrian in Cambridge, but that was before my time. Granit had worked in the same field as William Rushton who was my supervisor as an undergraduate. They both had dinner parties.

AM: Did it change your life? Obviously, there is some money attached. Presumably this helped your Royal Society professorship later.
AH: The Prize money helped us to have our second family.

AM: You may have put on some weight as well.
AH: I did not put on as much weight as I did when I was President of the Royal Society.

AM: Tell me about the Royal Society – that is another role – how long were you President for?
AH: I was President of the Royal Society for five years. Age limits to all sorts of appointments came in with new statutes in 1926. J.J. Thompson had become Master of Trinity in 1918 and at that time there were no age limits. Under the new statutes, those already in post, were allowed to remain under the old statutes and J.J. Thompson did. He remained Master until he died in 1940 at the age of 84. This meant that he was a Master when I was an undergraduate. I was an Entrance Scholar so I did come to some of his lunches.

AM: What was he like?
AH: He was very conversable and had a great interest in cricket which I am afraid I am ignorant about. His wife and daughter were staying with him at the time so I met both of them. He was absent minded. There is a story about him though I do not if it is true. One morning he went out to Cambridge and his wife found his trousers lying on the chair in their bedroom. She assumed he had gone out without his trousers and went out to catch him in the streets of Cambridge.

AM: Did she? Or was there another explanation that he had another pair of trousers, well, we shall never know! It is like Alan Smith, the economists, who is reputed to have often gone off in his pajamas to wander up and down the Fife coast. Absentmindedness in academic life seems to have some affinity.

AH: I am becoming more and more absentminded, but I think it is a matter of age.

AM: Another person you mentioned you met once, I think, or may be more, was John Maynard Keynes.
AH: I had an introduction to John Maynard Keynes from my half-brother Aldous Huxley. I was invited to a lunch in my first term. We did not keep up the acquaintance, but I got to know his nephews quite closely, particularly, Richard Keynes. He had been at school at Oundle with Geoffrey Trevelyan. He was a few years younger than me, but he came up to Trinity just before the war and Geoffrey Trevelyan introduced him to me. His [Keynes'] work has been close to mine. He applied the use of radioactive tracers to measure the movements of ions in and out of nerve and muscle fibres when they were active. He also obtained direct proof of some of the idea that had been originally rather hypothetical when we got onto it first. His wife was the daughter of First Lord Adrian. She was a very good singer.

AM: They had two pianos in their living room, I noticed.
AH: I do not think I played duets with either of them.

AM: Sometimes it is an impossible question, but was there during the whole of your life, anyone, who really struck you as amazing that is sort of the stature of Newton and Einstein and really out of this world. I can think of one or two in my life in Cambridge, but who would you rate as very, very…? And where did his brilliance lie?
AH: The most brilliant person I knew was Alan Hodgkin. Quaker ancestry again. Hodgkin's disease was named after a great-uncle of his. He has a very readable autobiography. I recommend your read it if you have not. I think I am its ideal reader because until reading it I knew nothing about his boyhood and this has a lot in it. There is quite a lot about his time in Cambridge. I only slightly knew his friends. I did meet him a few times during the war. He was developing radar for the Air Force at Malvern while I was in operation research for Anti-Aircraft Command, also in Malvern. I found myself in the same place as him and I visited him a number of times.

A radar trace is a spot on the cathode ray tube that goes across very quickly perhaps at over a thousand times a second only taking a very small fraction of a thousandth of a second to go across. It is started when a pulse is sent out from the radar station and it shows a deflection when the reflection from the aircraft comes back. The strength of this reflection varies a great deal from second to second. Bernard Lovell, a very distinguished person, still alive, was wondering how to record these fluctuations and mentioned this to Alan [Hodgkin] and I had an idea of how to do it. I used a Leica camera, given to me on my 21st birthday, that I still use, to photograph the traces from a cathode ray tube. I got the necessary series of photographs by winding a whole cassette of film through with a cover on the lens and then disconnecting the ordinary wind-on process as has to be done for rewinding film and rewound at great speed by winding a tape round the wheel of the rewind nob and fix up the camera on a tripod facing the cathode ray tube. Then when the series of traces was going across, I would pull the tape and get a long series of these traces which fluctuated in amplitude with high frequency. This was very useful to Lovell and Hodgkin, particularly the people who were developing radar.

I had got interested in Anti-Aircraft Command, again through A.V. Hill. During the First World War, he had been in a team developing anti-aircraft gunnery at Portsmouth and was, therefore, in touch with General Pile who was in charge of Anti-Aircraft Command. Pile needed a scientific advisor and Hill put him in touch with Patrick Blackett, the physicist, another Nobel Prize winner and appointed Blackett as scientific advisor. Blackett, naturally, wanted some dogsbodies, so A.V. Hill provided him with three assistants, one was Leonard Baylis, then a lecturer at University College, one was A.V. Hill's own son, David, and the third through David was myself. So, we were three physiologists in anti-aircraft gunnery. We had all had a very suitable background as physiologists. We had all taken the Tripos at Cambridge and had all done physics and a half subject of mathematics in Part One, that gave us the necessary knowledge of statistics and of differential equations to deal with the problems of anti-aircraft gunnery. Our work was largely a matter of adapting the predictors, the things that told the guns where to point, to use radar data. The existing predictors all had a pair of telescopes on them which were operated by two operators – one in elevation and the other in rotation in azimuth.[41] But they required data that was very precise to work satisfactorily. At that time the radars that were available were much too imprecise for the predictors to use the data correctly. So, we invented modifications to the predictors that enable them to make use rather indirectly of this very poor input data. We were working in an outstation of an Anti-Aircraft Command.

Then, Blackett was transferred first to Coastal Command and then Admiralty. We were for a time under Neville Mott, later Master of Caius and another distinguished physicist, and finally under Basil Schonland, a South African physicist, who had worked mostly on lightning and how it is produced. It was an interesting time. In late 1941, Blackett got me transferred to the Admiralty where I did the same sort of work in operational research in the gunnery division of the Naval Staff. The gunnery division was not concerned with the design of guns – that was under the controller of the Navy, the Third Sea Lord. The gunnery division was concerned with the use of guns and their accuracy. I had several trips on warships, the King George V, one of the first battleships during trials of its guns, shooting at one of the skerries, small rocks off the north of Scotland, based in Orkney [Islands], in Scapa Flow. Then I had a time in the Mediterranean on board the Belfast – a six-inch cruiser – which is now anchored in the Thames.

AM: I talked to a number of distinguished scientists like Patrick Bateson and Gabriel Horn and others about whether there are features of Cambridge, the organization of the university that provide good conditions for scientists and others to work in. They seemed to think that was the case and that was one of the reasons why they had not gone to America. Have you found it a good place to work and for what reasons?

AH: Cambridge is a very good place to work, particularly in my field which has a strong tradition in Trinity. Adrian was one the fairly early people and then Rushton, my supervisor, learned from Adrian and Alan Hodgkin, my mentor, learned from Rushton. Richard

Keynes, learned from Hodgkin, under whose supervision he took his PhD. It is largely a matter of chance. All my near relatives went to Oxford. I had one full brother, we were both at Westminster School, which has links going back to Queen Elizabeth I, with Christ Church Oxford and with Trinity in Cambridge. My brother got a closed scholarship to Christ Church and I got a closed exhibition to Trinity. Cambridge was the place to come to for people with an interest in science. My brother got a scholarship in history. My mother felt it would be better for me at a different place than my brother and that is how it turned out. How traditions develop is hard to say, clearly a large element of chance, people tend to argue how much is chance and how much is cause and effect and which is it. My answer is always that it is a bit of both.

AM: Random variation and selective retention as someone….
AH: Yes, process analogous to Darwinian natural selection.

AM: What relation is Francis. I knew an anthropologist called Francis Huxley.
AH: Francis Huxley is the younger son of my half-brother Julian Huxley. He is a half-nephew. He is only a few years younger than me. He has been in the States for a long time, I have not seen him for many years.

AM: So, perhaps on that family note of a great distinguished Cambridge family I would like to thank you very much indeed.
AH: Well, thank you it has been interesting recalling these memories.

AM: Have you written any of them down anywhere?
AH: Alan Hodgkin said that I ought to write them down. I have got about 40 pages of typescript intended for internal consumption. Whether it will turn out to be publishable or not, I do not know. Alan Hodgkin told me about his autobiography where he really started dealing with his war work because some of his colleagues in developing radar had been killed in flying accidents, testing the radar. He started it, essentially to make their work known and then extended it in detail with his earlier life, his time as an undergraduate, but only a little about his time as President of the Royal Society and Master of Trinity.

AM: What relation is Dorothy Hodgkin? I imagine she's a relation.
AH: Originally the wife of a first cousin of Alan Hodgkin's.

AM: Quite distant.
AH: Fairly distant.

AM: Thank you very much indeed.

NOTES

1. Andrew Huxley. 2021. Wikipedia. https://en.wikipedia.org/wiki/Andrew_Huxley
2. Thomas Henry Huxley, see Appendix.
3. Matthew Arnold, see Appendix.
4. Thomas Arnold, see Appendix.
5. In Greek mythology, an Oread or Orestiad is a mountain nymph. They differ from each other according to their dwelling: the Idaeae were from Mount Ida, Peliades from Mount Pelion, etc. They were associated with Artemis, since the goddess, when she went out hunting, preferred mountains and rocky precipices. The term itself appears to be Hellenistic. https://en.wikipedia.org/wiki/Oread.
6. Aldous Leonard Huxley, see Appendix.
7. Sir Julian Sorell Huxley, see Appendix.
8. Huxley, Thomas Henry. 1909. *Autobiography and Selected Essays*. https://bit.ly/2XwFkXZ
9. David Bruce Huxley, see Appendix.
10. Martin John Spencer Rudwick, see Appendix.
11. Sir Thomas Lewis, see Appendix.
12. Alan Lloyd Hodgkin, see Appendix.
13. Makoto Endo worked under Professor Andrew Huxley from September 1962 to April 1964 at the Department of Physiology, University College London. For more, see ams.org/notices/201305/rnoti-p576.pdf & https://bit.ly/3xPlVhn
14. *Hiragana* is a Japanese syllabary, one component of the Japanese writing system, along with *katakana*, *kanji* and in some cases Latin script. It is a phonetic lettering system. The word *hiragana* literally means 'ordinary' or 'simple' *kana* ('simple' originally as contrasted with *kanji*). https://en.wikipedia.org/wiki/Hiragana
15. *Katakana* is a Japanese syllabary, one component of the Japanese writing system along with *hiragana*, *kanji* and in some cases the Latin

script (known as *rōmaji*). The word *katakana* means 'fragmentary *kana*', as the *katakana* characters are derived from components or fragments of more complex *kanji*. *Katakana* and *hiragana* are both *kana* systems. With one or two minor exceptions, each syllable (strictly mora) in the Japanese language is represented by one character or *kana*, in each system. Each *kana* represents either a vowel such as 'a' (*katakana* ア); a consonant followed by a vowel such as 'ka' (*katakana* カ); or 'n' (*katakana* ン), a nasal sonorant which, depending on the context, sounds either like English m, n or ng ([ŋ]) or like the nasal vowels of Portuguese or Galician. https://en.wikipedia.org/wiki/Katakana

16. Stevn Runicman, see Appendix.

17. Johann Sebastian Bach (1685–1750) was a German composer and musician of the late Baroque period. He is known for instrumental compositions such as the Cello Suites and *Brandenburg Concertos*; keyboard works such as the *Goldberg Variations*, *The Well-Tempered Clavier* and the Toccata and Fugue in D minor; and vocal music such as the *St Matthew Passion* and the Mass in B minor. Since the 19th-century Bach Revival, he has been generally regarded as one of the greatest composers in the history of Western music. https://en.wikipedia.org/wiki/Johann_Sebastian_Bach

18. Franz Peter Schubert (1797–1828) was an Austrian composer of the late Classical and early Romantic eras. Despite his short lifetime, Schubert left behind a vast oeuvre, including more than 600 secular vocal works (mainly lieder), seven complete symphonies, sacred music, operas, incidental music and a large body of piano and chamber music. His major works include 'Erlkönig' (D. 328), the Piano Quintet in A major, D. 667 (*Trout Quintet*), the Symphony No. 8 in B minor, D. 759 (*Unfinished Symphony*), the 'Great' Symphony No. 9 in C major, D. 944, the String Quintet (D. 956), the three last piano sonatas (D. 958–960), the opera *Fierrabras* (D. 796), the incidental music to the play *Rosamunde* (D. 797), and the song cycles *Die schöne Müllerin* (D. 795) and *Winterreise*. https://en.wikipedia.org/wiki/Franz_Schubert

19. Johannes Brahms (1833–97) was a German composer, pianist, and conductor of the Romantic period. Born in Hamburg into a Lutheran family, he spent much of his professional life in Vienna. He is sometimes grouped with Johann Sebastian Bach and Ludwig van Beethoven as one of the 'Three Bs' of music, a comment originally made by the nineteenth-century conductor Hans von Bülow. https://en.wikipedia.org/wiki/Johannes_Brahms

20. Ludwig van Beethoven (1770–1827) was a German composer and pianist. Beethoven remains one of the most admired composers in the history

of Western music; his works rank amongst the most performed of the classical music repertoire, and span the transition from the classical period to the romantic era in classical music. His career has conventionally been divided into early, middle, and late periods. The 'early' period, during which he forged his craft, is typically considered to have lasted until 1802. From 1802 to around 1812, his 'middle' period showed an individual development from the 'classical' styles of Joseph Haydn and Wolfgang Amadeus Mozart, and is sometimes characterized as 'heroic'. https://en.wikipedia.org/wiki/Ludwig_van_Beethoven

21. Benedict Delisle Burns, FRS (1915–2001) was a neurophysiologist and a pioneer of operations research and of the statistical analysis of neuronal activity. He was elected Fellow of the Royal Society in 1968. https://en.wikipedia.org/wiki/Benedict_Delisle_Burns

22. William Albert Hugh Rushton, FRS (1901–80) was professor of Physiology at Trinity College, Cambridge. His main interest lay in colour vision and his Principle of Univariance is of seminal importance in the study of perception. https://en.wikipedia.org/wiki/W._A._H._Rushton

23. Edgar Douglas Adrian, 1st Baron Adrian, see Appendix.

24. Horace Barlow, see Part 3, p.

25. Lipid bilayer: The lipid bilayer (or phospholipid bilayer) is a thin polar membrane made of two layers of lipid molecules. These membranes are flat sheets that form a continuous barrier around all cells. The cell membranes of almost all organisms and many viruses are made of a lipid bilayer, as are the nuclear membrane surrounding the cell nucleus, and membranes of the membrane-bound organelles in the cell. The lipid bilayer is the barrier that keeps ions, proteins and other molecules where they are needed and prevents them from diffusing into areas where they should not be. Lipid bilayers are ideally suited to this role, even though they are only a few nanometers in width, because they are impermeable to most water-soluble (hydrophilic) molecules. Bilayers are particularly impermeable to ions, which allows cells to regulate salt concentrations and pH by transporting ions across their membranes using proteins called ion pumps. https://en.wikipedia.org/wiki/Lipid_bilayer

26. Alexander von Muralt, see Appendix.

27. Robert Stämpfli, see Appendix.

28. Sarcomere: Following the I-band is the A-band (for anisotropic). Named for their properties under a polarized light microscope. An A-band contains the entire length of a single thick filament. The anisotropic band contains both thick and thin filaments. Within the A-band is a paler region

called the H-zone (from the German 'heller', brighter). Named for their lighter appearance under a polarization microscope. https://en.wikipedia.org/wiki/Sarcomere

29. Isotropic bands: In physiology, isotropic bands (better known as I bands) are the lighter bands of skeletal muscle cells (a.k.a. muscle fibres). Isotropic bands contain only actin-containing thin filaments. The darker bands are called anisotropic bands (A bands). Together the I-bands and A-bands contribute to the striated appearance of skeletal muscle. Isotropic bands indicate the behaviour of polarized light as it passes through I bands. https://en.wikipedia.org/wiki/Isotropic_bands

30. Kenneth S. 'Kacy' Cole, see Appendix.

31. J.Z. Young, see Appendix.

32. The Hodgkin–Huxley model, or conductance-based model, is a mathematical model that describes how action potentials in neurons are initiated and propagated. It is a set of nonlinear differential equations that approximates the electrical characteristics of excitable cells such as neurons and cardiac myocytes. It is a continuous-time dynamical system. Alan Hodgkin and Andrew Huxley described the model in 1952 to explain the ionic mechanisms underlying the initiation and propagation of action potentials in the squid giant axon. They received the 1963 Nobel Prize in Physiology or Medicine for this work. https://en.wikipedia.org/wiki/Hodgkin–Huxley_model

33. The Regius Professorship of Physic is one of the oldest professorships at the University of Cambridge, founded by Henry VIII in 1540. 'Physic' is an old word for medicine (and the root of the word physician), not *physics*. https://en.wikipedia.org/wiki/Regius_Professor_of_Physic_(Cambridge)

34. John Alfred Ryle, see Appendix.

35. Guy's Hospital is an NHS hospital in the borough of Southwark in central London. It is part of Guy's and St Thomas' NHS Foundation Trust and one of the institutions that comprise the King's Health Partners, an academic health science centre. It is a large teaching hospital and is, with St Thomas' Hospital and King's College Hospital, the location of King's College London GKT School of Medical Education. https://en.wikipedia.org/wiki/Guy's_Hospital

36. Addenbrooke's Hospital is an internationally renowned teaching hospital and research centre in Cambridge, England, with strong affiliations to the University of Cambridge. Addenbrooke's Hospital is based on the Cambridge Biomedical Campus. The hospital is run by Cambridge University Hospitals NHS Foundation Trust and is a designated academic health science

centre. It is also the East of England's major trauma centre and was the first such centre to be operational in the United Kingdom. https://en.wikipedia.org/wiki/Addenbrooke's_Hospital

37. Robert McCance, see Appendix.

38. Elsie Widdowson, see Appendix.

39. George Macaulay Trevelyan, OM, CBE, FRS, FBA (1876–1962) was a British historian and academic. He was a Fellow of Trinity College, Cambridge, from 1898 to 1903. He then spent more than twenty years as a full-time author. He returned to the University of Cambridge and was Regius Professor of History from 1927 to 1943. He served as Master of Trinity College from 1940 to 1951. In retirement, he was Chancellor of Durham University. https://en.wikipedia.org/wiki/G._M._Trevelyan

40. Sir Charles Philips Trevelyan, 3rd Baronet (1870–1958) was a British Liberal Party, and later Labour Party, politician and landowner. He served as President of the Board of Education in 1924 and between 1929 and 1931 in the first two Labour administrations of Ramsay MacDonald. https://en.wikipedia.org/wiki/Sir_Charles_Trevelyan,_3rd_Baronet

41. azimuth

PART two

Richard Keynes. Photograph courtesy Alan Macfarlane.

Richard Keynes

Richard Darwin Keynes, CBE, FRS, (1919–2010) was a British physiologist. Keynes was the eldest son of Geoffrey Keynes and his wife Margaret Elizabeth (née Darwin), daughter of George Darwin. He was educated at Oundle School before going up to Trinity College, Cambridge.

During the war, Keynes served as a temporary experimental officer at the Anti-Submarine Establishment and Admiralty Signals Establishment (1940–45), returning to Cambridge after the war to complete his degree (1st Class, Natural Science Tripos Part II, 1946). Keynes remained at Trinity College as a Research Fellow between 1948 and 1952, winning the Gedge Prize in 1948 and the Rolleston Memorial Prize in 1950. During his career at Cambridge, he was a Fellow of Peterhouse (1952–60, and an Honorary Fellow, 1989); Head of the Physiology Department, and first Deputy Director (1960–64), then Director (1965–73); Director of the ARC Institute of Animal Physiology (1965–72); Professor of Physiology (1973–87); Fellow of Churchill College, since 1961.

Outside Cambridge, Keynes's positions included: Secretary-General of the International Union for Pure and Applied Biophysics (1972–78), then Vice-President (1978–81) and President (1981–84); chairman of the International Cell Research Organisation (1981–83) and the ICSU/Unesco International Biosciences Networks (1982–93); President of the European Federation of Physiological Societies (1991); a Vice-President of the Royal Society (1965–68); Croonian Lecturer (1983); Fellow of Eton College (1963–78); foreign member of the Royal Danish Academy (1971), American Philosophical Society (1977), American Academy of Arts and Sciences (1978) and the American Physiological Society (1994).

He was the editor of The Beagle Record: Selections From the Original Pictorial Records and Written Accounts of the Voyage of the H.M.S. Beagle *(1979) and* Charles Darwin's Zoology Notes and Specimen Lists from H.M.S. Beagle *(2000). His other publications, written alongside other authors, include* Nerve and Muscle *(2001) and* Comparative Physiology of Sensory Systems *(1984).*[1]

I

Alan Macfarlane (AM): It's a very great pleasure and it seems like a long time to have been in Cambridge without having met Richard Keynes, so thank you very much indeed for agreeing to be interviewed today. Richard, when were you born?
Richard Keynes (RK): 14 August 1919.

AM: Where was that?
RK: In London.

AM: Usually I ask people about their parents but in your case can we start at your great-grandparents. Who were your great-grandparents?
RK: My great-grandparents – the best known one was Charles Darwin.[2] That picture up on the wall is the family copy of him taken in his old age by my uncle, Leonard Darwin.[3]

AM: Darwin has obviously had a huge influence on your life.
RK: Yes, but the story about that is that actually in 1968, I had still never really taken a particular interest in Darwin which always rather hurt my mother who was a big Darwinist. But I simply had not taken a great interest in him. I was then travelling in South America, from working with a friend in Chile back through Buenos Aires. People in Buenos Aires said, 'Have you seen the Darwin collection in Buenos Aires?' and I said no. But then it turned out that the Argentinian professor who was showing me around knew of this collection and he took me to see it. These belonged to a man called [Eduardo] Braun-Menéndez[4] who I think was a descendant of one of the wicked Germans who used to own most of Patagonia but was perfectly charming, a lawyer actually. He had in his possession two sketch books painted by Conrad Martens[5] who was the second official painter on the dawn voyage of *The Beagle*. These all are

This interview was conducted by Alan Macfarlane on 26 September 2007. The transcription for this interview has been provided by Rittika Dhar, for this volume.
These interviews are available for viewing on:
https://www.sms.cam.ac.uk/media/1124175
https://www.youtube.com/watch?v=vMDjaD-9fbU&t

now in the University Library because I arranged for Nora Barlow[6] to – she had a convenient fund – and he asked me to sell them. I could not understand why he wanted to sell them. It did not seem to be because he was not rich enough. We left it at that and then I thought it would be much better, knowing that Nora had this fund which she used to use for giving things to Down House and Darwin connected things, I thought she would pay for it. I got Christie's to value these things which they were then worth gigantically less than they are today. They were bought then and those pictures are now in the University library.

AM: After having discovered this Darwin collection in Buenos Aires, when you came back did you start investigating?
RK: When I came back, I started investigating. I eventually published a book called *The Beagle Record* which has a lot of the pictures which were in that book reproduced in it. That was the first big book that I produced. It was all mostly quotations from family letters to Darwin during the voyage of *The Beagle*.

AM: Were these family letters already in the University library?
RK: Yes, they were. I did not have to do any research to get them. They were all there.

AM: But it is interesting that he did not really have an influence on your academic career at all except for getting interested later. It did not turn you into becoming that academic that you did.
RK: No, the person who did that was A.V. Hill,[7] who was Polly Hill's father. I knew him extremely. It was certainly him who resulted in my being a physiologist.

AM: Was it? He was a Nobel Prize winner, wasn't he?
RK: He was a Nobel Prize winner.

AM: And a physiologist?
RK: Yes.

AM: And how was he related to you?
RK: He was my uncle. Yes, that is right. And I knew him extremely well. I went on knowing him till he died, when he was fairly old anyway.

AM: What sort of person was he? Autocratic?
RK: Polly would have said so, yes.

AM: That's why I said it.
RK: I do not think his make relatives would have ever found him that.

AM: There is a nice portrait of him in King's College which I look at. We have moved down to the next generation now. So, we have got the Hills on one side and the Keyneses. How did the Keyneses come in?
RK: My father was a Keynes. The portrait up in the room next door is him because he was the first University Registrary and a Fellow of Pembroke. Then he got thrown out of – he was a sort of economist – his college Fellowship when he did such wicked things, going to be a Registrary. I think Pembroke made him an honorary Fellow later on but I know there was a sort of faint feeling against the college sometimes during my childhood for that reason.

AM: Registrary is an administrative officer. Was he also an academic or book collector?
RK: Yes, he was, he was. He was a perfectly sound academic.

AM: In what field?
RK: Well, it was not economics. It was....

AM: Statistics?
RK: Yes, that sort of thing. He was a dear old man and I remember him quite well though we did not talk very much when he was a very old man. He died in his 90s.

AM: Was this your father?
RK: My father's father.

AM: Grandfather. Of course, still two generations back. And do you remember your grandmother on that side?
RK: Extremely well, because she was the second woman Mayor of Cambridge.

AM: What was her name?
RK: Frances Ada Keynes. She was a very, very good administrator obviously and an absolutely splendid person. She ran everyone but in a most unbossy sort of way.

AM: You spent a lot of time with your grandparents, did you? Or did you not?
RK: I visited them very frequently.

AM: Because they were living in Cambridge.
RK: My other grandmother, my American grandmother, lived in Cambridge too and she didn't die until after the Second World War. I would go and stay with her quite a lot during the War.

AM: How are you related to Gwen Raverat[8] then?
RK: She was my aunt, my mother's sister.

AM: So, *Period Piece*, which we discussed off-camera was about your childhood as well?
RK: No, it was not because the reason that *Period Piece* got written was that when my American grandmother came over and got married, she then after that wrote a lot of letters to her family at home in Philadelphia which contained all sorts of details about social existence in Cambridge which no English wife like this would have thought would have any interest to people. But she wrote an absolutely splendid collection of letters. I do not know what happened to those because my mother and Gwen thought maybe they ought to be published, they were so good. But then it turned out there were probably so many people still alive who had been children when *Period Piece w*as coming into life and it might be embarrassing. Then Gwen said, 'Oh, I'll write a book about it all'. She sat down and wrote *Period Piece* with a lot more of her own drawings and things to illustrate it. And it is still in print.

AM: Yes, and it is still a bestseller. It is a lovely book. What was she like as a person? Do you remember her?
RK: I liked her very much, yes. She was quite fierce sometimes.

AM: Well, she was frightened by those tigers under the bed, I seem to remember, when she was a child. She had to be fierce. We are moving down to that generation, your parents' generation. Tell me something about your mother and father.
RK: My mother was a very good mother, very good indeed. She did hospital things. My father was a senior surgeon at Bart's until 1960. After that, he turned into writing books by the dozen, bibliographies.

AM: What was his Christian name, remind me.
RK: Geoffrey.

AM: Geoffrey. Ah yes, the famous bibliographer.
RK: But he died in 1993, still writing another book. I do not think I can equal him.

AM: In age or in books?
RK: Certainly not in books, but I have read quite a lot of books of his.

AM: But she did not have an academic career really? She was mainly in the home?
RK: No, she was at Oxford. She was at Somerville in Oxford. But I do not remember that.

AM: And it was your father's brother who was John Maynard,[9] is that right?
RK: That's right.

AM: Yes. How do you remember him, in what way do you remember him?
RK: I remember him very kindly because he had a reputation for being rather fierce sometimes but he was always very kind to young nephews, very kind indeed.

AM: Do you remember any particular incidents in your relationship, any book he suggested?
RK: He made one mistake because he made me – as a present to the family, a Fellow of the zoo. I was famous because I had known all the trees backwards when we went to the zoo in London. At the beginning of the war, he gave up since we obviously would not be able to do it. But what it never occurred to him to do was to make me a Life Fellow because if he had done, I could have gone on with no further appointments for the rest of my life. I thought that was the kind of thing he ought to have worked out for himself.

AM: As an economist, it would have been much cheaper.
RK: But he was very nice indeed.

AM: Did you ever see him working because one of the things that struck me about the portraits of him in King's College is that the

way he managed to produce so many books on so many different subjects may have been related to the fact that he had this – in the portraits he would be sitting with a sort of green wooden shelf or tray on his knee in a sofa or on a chair with a little hat on, scribbling away. I had the sense that, as many busy men are, he was very good at organizing his time and could work under any circumstances.
RK: I think that is probably true. It must be true, yes, because he certainly worked continuously and he was always writing his books. But the great character there that I loved so much was my aunt Lydia, who was the Russian.

AM: Lydia Lopokova?[10] Yes.
RK: Lydia Lopokova. I became her, along with Lord Kahn[11] – two executives, trustees. She, after all, lived on for another twenty something years after Maynard died. I was in charge of her personal things and her money was looked after by Richard Kahn.

AM: That's why you edited her letters to John Maynard.
RK: We inherited her things. Her descendants included us, Polly Hill and I. We thought that we ought to publish her letters so Polly and I actually edited those letters. The originals are all back in King's now. We used to know Lydia extremely well. She was a wonderful character.

AM: Was she? You can see why Keynes married her, can you?
RK: You can indeed.

AM: What was special about her? She was just very warm?
RK: Lovely sense of humour and very warm.

AM: Had Keynes been very interested in the arts – he helped set up the arts theatre – before he married her?
RK: I think so, yes.

AM: But she also encouraged that interest?
RK: Sort of, yes. No, she was the most adorable person. She went on living in their house that he had, Tilton in Sussex. We used to go and take her to Glyndebourne every summer and go and stay with her for a little while. She was a very great character.

AM: When I first came to Cambridge, I was in the flat that came shared with her just above David's bookshop.
RK: That's right. That is where she was. She went on using that. It belonged to King's. She went on using that. Then she went to live permanently at Tilton. We used to have summer holidays at Tilton.

AM: Wonderful. One could have a whole session about your relatives but we ought to get down to you now. You were born in London, 1919 and do you remember anything about your first schools?
RK: Well, my first school was The Hall School, Hampstead, the boys with the little red cross on their caps. That was a very good school I thought. It was down by Fifth Row, the big crossing at the bottom of the hill and was a very good school.

AM: This was just when you were seven or eight? Was it a preparatory school?
RK: Yes, a preparatory school. I must have started at seven or eight and then went on till I went to Oundle when I was thirteen probably.

AM: Yes. There was probably no one at your first school who you particularly remember influencing you or setting you in any direction.
RK: No, I do not think so. I do not think that.

AM: I do remember, I went to the Dragon School at Oxford and there are several teachers now who I still think probably had some influence on me, a communist who was an improbable member of the staff, a great Gilbert and Sullivan fan I became.
RK: My Keynes grandfather was a great Gilbert and Sullivan fan. He had a unique collection of records.

AM: Going on to Oundle then, how did you find that?
RK: I was very happy there.

AM: All the time?
RK: Yes, I think so. I was rather glad that I had been there in what I am not sure that they still do, which is that everyone in the school spends one week every term working in the workshops.

AM: What kind of workshops?
RK: Wood workshops, metal workshops.

AM: I am sure they do not do that.
RK: If you were a scientist, the week you might have done in the workshops you instead just spent doing experiments in the school. It was very successful. That thing, it must have been introduced by that famous headmaster. I had a very good housemaster, Ian Hepburn.[12] He was a chemistry master but he was a very, very good housemaster.

AM: Were you head of house?
RK: No. One of my very closest friends was always the head of the house and was a much better head than I would have been. He ended up by getting three MCs [Military Crosses] in the war. If you could think of anyone that you would have liked to have as your leader in the infantry, it would have been him.

AM: That sounds a familiar name. Did he go on to do anything after the war?
RK: Not particularly. He became a businessman. He lived in Newcastle so he belonged to the appropriate people there.

AM: This housemaster influenced you. Were there any other teachers who particularly influenced you?
RK: One other one I remember I forget his name.

AM: Don't worry but what did he do?
RK: He taught French. He was Victor Rothschild's secretary for a long time after that and he always used to run the school plays. I was in two of the school plays and then there was one day when the masters' wives, who were taking the female parts, were all not there. Our master said, 'I'll do that'. He got us all in total hysterics in no time at all. He was a good person.

AM: But he did not encourage you to do French?
RK: Well, no. But I had been having holidays in France. The important thing that I was able to do in France which I wish we had been able to do for my own children was to go to the Abbaye de Pontigny, which was in Burgundy. They had a famous place for where the socialists used to go to meet and they had an international summer school during the summer. I spent three months there in Burgundy. Most lovely place, you would hardly get sleep in the night because of nightingales. Everyone was very nice there. We all had to

speak French because there was only one other Englishman, whom I still meet actually. But otherwise, they were all different nationalities so we all had to speak French.

AM: How old were you at this time? You were still at school?
RK: I was 18 or 17.

AM: Were you beginning to feel that science was what you wanted to do at this point?
RK: No, A.V. Hill had me firmly on those lines a long time earlier. His sons were all scientists.

AM: There was no question that you were going to do – was he specific that you were going to be a physiologist?
RK: Well, he was a physiologist. I think it was.

AM: And you did reasonably well and got all your exams. Did you have any other hobbies, particularly? You were in the school play. What were the plays? Were they [William] Shakespeare?[13]
RK: Oh no. They were well-known plays. But I cannot remember them.

AM: Did you have any particular hobbies at that time apart from learning French and listening to nightingales?
RK: I eventually became an ornithologist but that was more because I was sent on the public schools' exploring society to a trip right up to the north of Norway. For some reason I said I was interested when I was being interviewed by those who ran the place. I did not know much about it actually. But then when I was at Oundle, the Headmaster, Kenneth Fisher, always used to have a boy to carry for him, carry a telescope that was used for bird-watching. I became a great pal of Fisher's actually. He was best to teach German. But I must say I didn't think he was a very good teacher of German. He was a quite successful Headmaster though.

AM: When you finished at Oundle, did you try for several universities or was it just Cambridge?
RK: No, just Cambridge. All my relations had been in Cambridge.

AM: And did you try several colleges or just – you went to Trinity?
RK: Just Trinity. And A.V. pushed me to do physiology then. I think he subsequently examined me for my PhD.

AM: Really? That is very incestuous.
RK: Very incestuous.

AM: But as an undergraduate, you did physiology? You did natural science?
RK: I did natural science.

AM: And you got a first, did you? Or was it in two parts anyway?
RK: No, I did not. Well, I did later on. But the War came. When I went to Oundle, I was in junior house and so after the end of my first year, I was allowed to take the entrance examination to Oundle and astonished everyone by coming top in everything. But when the wars started, that was in my second year....

AM: As an undergraduate.
RK: I went to do physics first of all. But after two terms, I decided that if I stayed on and took the tripos, having only done one of the Tripos in physics, I might have failed altogether which wouldn't have done me any good at all. I knew and A.V. knew – he was right in the thick of it and organizing all the sciences around the place – I then switched and then I just went to work for the Admiralty.

AM: That's right. Yes. You worked on submarines or related to submarines.
RK: I first of all worked on submarines, yes. That is a long story that I will not go into.

AM: Can you abbreviate it?
RK: I put it all in the book. The most useful thing I did during the war was that I was working at Fairlie. We had been moved out of Weymouth dockyard up to the Clyde because we needed to have submarines in the neighbourhood to do experiments on them. We were in Scotland. Then in 1942, which was the year when the number of ships torpedoed got to a top, got really bad, I found that everybody at Fairlie was absolutely fed up with the Director. It was his attitude that really prevented us from doing any work. But the senior members of the staff, who were perfectly willing to talk to me, one of the most junior members of the staff, also felt strongly that they ought to be doing something rather different. They were not in

a position that they could go to the director of sound research in the Admiralty and complain because they were really on the staff. But there wasn't anything that I and one of my friends, Roger West – we could be thrown out but we could not be punished and we had all signed the secret papers. I finally decided that we could really do something about it. I wrote a long memorandum on what was wrong with the institute and after a fierce argument with the Director, he agreed to send it to the Admiralty. Other people had complained before but the Admiralty in London had managed to squash it all, which was quite wrong. Then I did a beautiful controlled leak and sent a copy of my long report to my cousin, Maurice Hill, who was at home. He immediately gave it to A.V. and A.V. immediately went to the Athenaeum and had lunch with the Third Sea Lord and told him there was something going wrong in Fairlie which he better put right, which he did.

AM: They got rid of the Director?
RK: They sent him to be a British Representative in Washington where he didn't do any harm.

AM: Things looked up after that?
RK: Definitely. They elected – I cannot remember the name of the man, he was a physicist in Cambridge who was made Director instead of him. I never met him. I went to his lectures but I never met him. He did very well apparently, which does not surprise me at all.

AM: So, you may have saved our....
RK: I was useful. A.V. helped me very, very much in all that. I first of all met Patrick Blackett[14] who was then Director of Operational Research in the Admiralty. I made much more of an impression on Blackett than I had realized at the time because some good many years after that, I was on the council of the Royal Society when Blackett was President. Blackett was not the same person by all accounts. He had been wonderfully lively and an active man when he was much younger but he was rather old by then. He seemed much more impressed by me when I met him in 1942 than I realized, he would push me almost embarrassingly. But I had a good time going around the place with other people outside of that.

AM: So, you did your three or four years in war service?
RK: It was early 1940 to 1945 but I managed to get out. I got special quick release from the chief scientist and went and read Physiology Part II. Those people who came back to the University at that period were much better subsequently. They had all matured and they were better students to teach.

AM: This was about 1947 you got your Part II?
RK: 1946, I got my degree and then two years later I was made a Research Fellow of Trinity.

AM: What were you researching on?
RK: I was researching on nerves. My chief teacher was Alan Hodgkin[15] who was just a very great man, very well-deserved scientist and Nobel Prize winner. I worked with him longer than anyone else did, actually. I worked for about five years. He was an absolutely splendid person.

AM: I remember in my interview with Sydney Brenner,[16] he said that the only way a scientist really can learn – there's no other way other than by apprenticeship, working with a really good scientist in a lab.
RK: I think that is quite true. I was doing that. But when I started doing research, Alan Hodgkin and Andrew Huxley,[17] who also shared the Nobel Prize with him, were all busy doing modern electronics. I had been designing radar sets a good many years so I knew I bit of electronics. But that was a much easier time to start then because there were so many absolutely vital problems sitting on the desk in a way which they are just not there in this day. There are plenty of big institutes with hundreds of people doing research but they were all doing similar sorts of problems. It was much, much easier for us in 1948 than it would be today.

AM: The thesis you were doing, was it chosen for you?
RK: No, it was chosen for me by myself, I think. What I did was I started working on radio isotopes. It was a time when it had not been done. No one had used them in England for anything before. I had to start by discovering how to make radio sodium and potassium. Fortunately, there was an ancient machine for making radiators which had been laid up during the war and Cavendish recovered it

and so I was able to make my own isotopes and do quite a lot of things, pioneer things in that way.

AM: Where were you working? Were you in the Old Cavendish?
RK: No, I was working in the physiology department.

AM: Yes, that is over on the Downing side.
RK: Yes, yes.

AM: That's right, because I worked in the Old Cavendish laboratory in the rooms where they discovered pulsars, radioastronomy. And I know the physiology department. I had talked to Sydney about that.
RK: Yes, I bet you did. Sydney had likely rude names for all the most eminent people.

AM: He did not tell me those. I will have to have another session with him for the rude names. You were working on that and at the end of your research fellowship, what did you do?
RK: Well, it would have been nice to have stayed on as a teaching fellow but unfortunately, Trinity then had Hodgkin and Huxley and William Rushton,[18] who were all very eminent physiologists. They did not need another one. I was elected a Fellow of Peterhouse. Peterhouse was a very nice College because it was so much smaller.

AM: Yes. How long were you there?
RK: I was there for about five years – from 1952 to 1960 – and then I did something unacceptable to the then Master of Peterhouse.

AM: Who was that? Was that the archaeologist?
RK: No, no. He was a very good friend of mine. But it was his predecessor who I did not have a great deal of respect for because he had written a very important textbook on the history of science but never seemed to me the least interested in the history of science.

AM: Oh right, I think I know who you mean, the historian. He was a historian.
RK: He did not approve when I took a job as being Deputy Director and very soon afterwards Director of the Babraham Institute of Animal Physiology. I spent quite a few years there until I came back as Professor of Physiology.

AM: Why didn't he approve that? He thought it would take you away from your duties in Peterhouse?
RK: It had no connection with the University at that time. They now do. Since then, because I insisted on having been allowed graduate students there, they are now still in existence but they were run by the ARC, the agricultural research institute [Agricultural and Food Research Council (AFRC)]. But they do have some well-known people there and it is quite a good place.

AM: As you know, I know nothing about science and some of my questions may be naïve, but is there a distinction between animal physiology and human physiology? You are now running an animal physiology unit. Was that a deviation from previous work or just a continuation?
RK: No, they have both got a lot of the same things. I did not ever do any human physiology.

AM: You lost your Peterhouse Fellowship and quite soon became almost a Founding Fellow of Churchill. Were you a Founding Fellow?
RK: I was not counted as a Founding Fellow but I was in fact appointed one month or so after the real Founding Fellows. No, that was quite a happy time. We had some very good people in Babraham. But it seemed to me that most of the staff there were simply doing the same piece of research that they had got for getting their doctorates in the first place. They were not really finding anything.

AM: Do you feel, as I think I do, that it is very important to keep moving onto new areas and topics? Certainly, in the arts it is because after about five or eight years, the laws of diminishing returns set in and you become stale.
RK: I think that is quite true. Yes, I agree with that.

AM: Has your work been discovering new problems and moving on to them?
RK: Well, I am not sure it has much because I started working on squid giant nerve fibres with Hodgkin and went on doing that until I went to Babraham. And then when I went to Babraham, I could not really stop working on squid. That was work that could only be done for a short time in autumn when the squid were available so

I used to actually absent myself from being Director of Babraham and disappear to Plymouth.

AM: Is that where the squid were?
RK: I think I was the only member of one of these big institutes who had been sent for to London by the Director Elect too much for doing too much research myself. I did not have a great deal of respect for that particular Director.

AM: How did you enjoy Churchill, because recently I did an interview with another Founding Fellow, George Steiner.[19] You must know George?
RK: I know George very well, of course.

AM: He was also a good talker, isn't he? He obviously enjoyed association with Churchill very much. You have given all your papers to Churchill, haven't you? Your personal papers?
RK: I dumped a lot of stuff on them when I finally retired in 1986. I had a vast number of papers and nowhere to put them and no lab. I gave them all to Churchill archives and they can have the rest if mine later on. Maybe I will start sending them in now so I can have room for more books.

AM: You have already deposited 104 archive cases of papers. How many more do you think you have got?
RK: I should think the same number at least.

AM: Well, you were lucky to have an archive that would take all that.
RK: It is a very good archive. I was President of the archive committee at Churchill for totally indecent time. There was no rule that said that I wasn't allowed to be re-elected. I kept on being re-elected and when I finally said, 'Look, you've had enough of me' and we investigated when I started signing the minute books, it turned out I had been signing for 24 years. I enjoyed that very much. By then I was writing books primarily.

AM: We'll come to those books in a minute but coming back to Cambridge, we are now in the early 60s. When did you become a Professor?
RK: After I left Babraham.

AM: After you left Babraham, you became Professor of Physiology. And Director of the department?
RK: Yes, yes. That has all been switched now. Physiology has been fused with anatomy and I am not sure I quite approve of everything that has been done on behalf of the physiologists.

AM: Was your time as professor and director a fruitful and happy time?
RK: Yes. Actually, at that time and quite soon after then, I was spending quite a long time – I cannot remember when – we had a big meeting in Cambridge to see whether there should be an international society for biophysics founded. Everyone was in favour of that except Hodgkin and Huxley, who did not seem to approve for some reason. I never quite understood why not. But in any case, an institute was set up and – it is still running – I became Secretary of it for a long time and was doing things and then I became President of it ultimately. That resulted in my becoming a member of ICSU, International Council of Scientific Unions, which is the international body for science. The thing that I particularly enjoyed is that I became Chairman of an UNESCO body. What we considered this committee should do was to support Third World people to teach themselves and not just pour money at them but get it for themselves to work. While I was Chairman of the committee, we had to go meet somewhere every year and I went on for about 10 years doing this.

AM: Which countries did you go to? All over the world?
RK: All over the world. What really happened originally was, in 1951, a man called Carlos Chagas,[20] who was a Brazilian and son of the inventor of Chagas Disease, came to Cambridge to find someone to be the Visiting Reader that year and wanted to get Alan Hodgkin. Alan was never a very good man for travelling and Andrew [Huxley] was busy getting married so they both turned it down. But I could see no reason why I should not accept the invitation so I did accept it. Then I went and worked for three months in Rio where I worked on what I knew about the latest technique which you needed for discovering how electric eels actually produce their

electricity. I did that pretty successfully with a bit of collaboration with a man called Martins Pereira in Rio and greatly enjoyed that. Then I went on and became partly instrumental in trying to get Third World biologists properly educated. The only difficulty was we never had enough money. We were promised millions of dollars by UNESCO [United Nations Educational, Scientific and Cultural Organization] sometimes but we never actually got it. The people I met could really have made good use of that money. But the Royal Society I rather disagreed with because the first person – every year we had meeting at the Royal Society about ICSU affairs and I had already spent some time with members of the executive committee at ICSU working out how much we might increase the money we asked for from the countries in order just to keep going. We were asking for more every year and we always started by being heavily ticked off by the Royal Society for wanting any more money. I must say I did not approve of the Royal Society for this.

AM: Is this why – I noticed in the curriculum vitae that you have various honours and awards from Brazil.
RK: Yes, that is why. I got some honorary degrees from Brazil. Those were actually ones that I really did work for.

AM: You have had a long-term association with Brazil then.
RK: Yes.

AM: And you admire Brazilian scientists?
RK: The ones I worked with were good, yes. In all those countries, sometimes they are very good, sometimes they are not very good. Brazil was a leading one. Chile was the best country while I was doing that work.

AM: You have done quite a lot of administrative jobs, heads of and deputy directors and chairing archive committees. How do you balance between research, teaching and administration? Did you find they conflicted or did they feed into each other? Did you enjoy administration? Some people find it tedious.
RK: I enjoyed running these societies but I did not particularly enjoy being Director of the Institute. I had to determine who was going to be promoted, I did not colossally admire the ARC actually.

AM: That's the Agricultural Research Council? They were funding?
RK: They were funding it. Then for a long time I was on the Executive Committee of the Plymouth laboratory because I went on working – even when I was working in Babraham, I went quite often in the autumn for a month or so to work on squid there – and ARC did not approve of that. I had to face American colleagues mostly who came and worked with me and that was fun. Between 1960 and 1986, I was very tired. I was doing much more. I was always going back to squid for a while and I was writing lots of books and things but I was not really doing much research. I did not have a lab in Cambridge and I wish I had had more – I simply slipped a bit but I do not think I was really responsible because I could never get any support. The only thing that matters to these departments nowadays is how much money they have got. They have to ask for money and the government, the Medical Research Council, always flatly refused to give me any money for Babraham. I thought they were wrong but there was not anything I could do about it. I tried hard.

AM: Did you teach much? Did you lecture?
RK: I lectured as Professor, yes.

AM: Did you enjoy lecturing?
RK: Yes, yes. It was quite useful when I was ill and went to Addenbrooke's, which I had to do a fair number of times. Quite a lot of people who had been to my lectures were there, they had never done any harm.

AM: What about supervising? Did you supervise PhD students at all?
RK: Well, yes. One of my most famous PhD students was Jared Diamond,[21] who wrote the book on....

AM: Really? Yes, I know. *The Third Chimpanzee.*
RK: He was the cleverest graduate student I ever had. I have got his last book somewhere. The University wanted to appoint him as Professor of Geology and he did want to come to Cambridge but of course, the education of his sons....

AM: He turned it down, didn't he?
RK: I know, but he could not possibly. What you get paid in Cambridge was totally different from what you get paid in UCLA.

AM: Yes. But he was a very clever man? Nice man? Did you like him?
RK: Yes. He did not always do things the way I thought they ought to be done earlier on but he certainly got much more human after he got married. He was very bright.

AM: Yes, I like his work.
RK: The thing that has been occupying me in recent years is writing more books about Darwin.[22] The most important one by far that I have done was actually – turned out that his zoology notes made on *The Beagle* had never been published, they had just been sitting in the University library but had not been published. I produced this whole volume. I spent four years, hard work actually. I really felt that I was talking to the old man sometimes because it was all in his hand. Everyone always thought the most difficult thing was the handwriting but it was not actually. After all, they could all read each other's handwritings.

AM: Do you spend much time at Down House? Do you go there often?
RK: No, not recently. I cannot walk at all far because I have got a badly arthritic back.

AM: I have got some of the stones. You know he has a thinking path and I have a thinking path in my garden with some of the stones. He was reputed to have picked up a stone each time he walked around and keep a pile. When you say you got to know him or sort of felt you were talking to him, did you form any impression of his character?
RK: He was a charming and friendly person. He always got on with everybody.

AM: Seems quite a hypochondriac at some points.
RK: Yes. His illness certainly was partly exaggerated. There are two doctors in Chile who just produced a new diagnosis for him which is something called – I cannot remember the term, but it was a disease of your innards.

AM: He probably did have something.
RK: But I admire him immensely. The Cambridge University Press originally published it at 95 pounds. It did not sell very well. But

by great good fortune, they then utilized it and I think now you can buy it at 30 pounds. The book is the same size as the original and they ought to be able to sell other copies.

AM: They have put online a lot of Darwin's correspondence, haven't they, the University library?
RK: Yes, I have given them a lot of stuff to add to that.

AM: Did you? They have not added it yet.
RK: No, they have added it.

AM: Oh, they have added it now.
RK: For a long time, the notes that he made hadn't been properly printed. My aunt, Nora Barlow, who was a very, very good zoologist herself, edited his bird notes, which have been published some time ago. That kept us all going. And she edited *The Beagle* diary. I re-edited *The Beagle* diary 50 years after Nora's version of it had been published. It needed to be reprinted. There were very few mistakes that she made, I must say.

AM: You are still obviously very active and that is a good point perhaps to end the interview.
RK: I have just written what certainly has to be my last book which is a history of research on animal electricity. I hope that is going to be published by the American Philosophical Society.

NOTES

1. Richard Keynes. 2022. Wikipedia. https://en.wikipedia.org/wiki/Richard_Keynes
2. Charles Darwin, see Appendix.
3. Leonard Darwin, see Appendix.
4. Eduardo Braun-Menéndez, see Appendix.
5. Conrad Martens (1801–78) was an English-born landscape painter active on HMS Beagle from 1833 to 1834. He arrived in Australia in 1835 and painted there until his death in 1878. https://en.wikipedia.org/wiki/Conrad_Martens
6. Nora Barlow, see Appendix.
7. A.V. Hill, see Appendix.

8. Gwendolen Mary 'Gwen' Raverat (née Darwin; 1885–1957) was an English wood engraver who was a founder member of the Society of Wood Engravers. Her memoir Period Piece was published in 1952. https://en.wikipedia.org/wiki/Gwen_Raverat

9. John Maynard Keynes, see Appendix.

10. Lydia Lopokova, Baroness Keynes (1891–1981) was a Russian ballerina famous during the early 20th century. Lopokova trained at the Imperial Ballet School. She toured with the Ballets Russes in 1910, and rejoined them in 1916 after an interlude in the United States. Lopokova married the renowned English economist John Maynard Keynes in 1925, and was also known as the Lady Keynes. She largely disappeared from public view after Keynes's death in 1946, and spent her remaining years in Sussex. https://en.wikipedia.org/wiki/Lydia_Lopokova

11. Richard Ferdinand Kahn, Baron Kahn, CBE, FBA (1905–89) was a British economist. Kahn was born in Hampstead into the orthodox Jewish family of Augustus Kahn, inspector of schools and former German schoolmaster, and Regina Schoyer. He was brought up in England and educated at St Paul's School, London. He attended King's College, Cambridge. Kahn took a 1st in Mathematics, Part I, at Cambridge, followed in 1927 by a 2nd in Physics in the Natural Sciences tripos. Taught economics by Gerald Shove and John Maynard Keynes from 1927 to 1928, he gained a 1st in Economics, Part II, in 1928. In 1930, he was elected a Fellow of King's College. https://en.wikipedia.org/wiki/Richard_Kahn,_Baron_Kahn

12. https://en.wikipedia.org/wiki/Ian_Hepburn

13. William Shakespeare, see Appendix.

14. Patrick Blackett, see Appendix.

15. Alan Hodgkin, see Appendix.

16. Sydney Brenner, see Appendix.

17. Andrew Huxley, see Part I, p. 3.

18. William Rushton, see Appendix.

19. Francis George Steiner, FBA (1929–2020) was a Franco-American literary critic, essayist, philosopher, novelist, and educator. He wrote extensively about the relationship between language, literature and society, and the impact of the Holocaust. An article in *The Guardian* described Steiner as a 'polyglot and polymath'. https://en.wikipedia.org/wiki/George_Steiner

20. Carlos Chagas, see Appendix.

21. Jared Mason Diamond (b. 1937) is an American geographer, historian, ornithologist, and author best known for his popular science books *The Third Chimpanzee* (1991); *Guns, Germs, and Steel* (1997, awarded a Pulitzer Prize); *Collapse* (2005), *The World Until Yesterday* (2012), and *Upheaval* (2019). Originally trained in biochemistry and physiology, Diamond is known for drawing from a variety of fields, including anthropology, ecology, geography, and evolutionary biology. He is a professor of geography at UCLA. https://en.wikipedia.org/wiki/Jared_Diamond

22. Charles Darwin, see Appendix.

PART three

Horace Barlow. Photograph courtesy Alan Macfarlane.

Horace Barlow

Horace Basil Barlow, FRS (1921–2020) was a British vision scientist. He was the son of the civil servant Sir Alan Barlow and his wife Lady Nora (née Darwin). Barlow was educated at Winchester College, and earned an M.D. at Harvard University in 1946.

In 1953, Barlow discovered that the frog brain has neurons which fire in response to specific visual stimuli. This was a precursor to the work of Hubel and Wiesel on visual receptive fields in the visual cortex. In 1961, Barlow wrote a seminal article where he asked what the computational aims of the visual system are. He concluded that one of the main aims of visual processing is the reduction of redundancy, which has been extended to the efficient coding hypothesis. While the brightnesses of neighbouring points in images are usually very similar, the retina reduces this redundancy. His work thus was central to the field of statistics of natural scenes that relates the statistics of images of real world scenes to the properties of the nervous system.

Barlow was a fellow of Trinity College, University of Cambridge. He was elected a Fellow of the Royal Society in 1969 and was awarded their Royal Medal in 1993. He received the 1993 Australia Prize (along with Peter Bishop and Vernon Mountcastle) for his research into the mechanisms of visual perception, and the 2009 Swartz Prize for Theoretical and Computational Neuroscience from the Society for Neuroscience. He was awarded the first Ken Nakayama Prize from the Vision Sciences Society in 2016.

Some of his published works include 'Possible principles underlying the transformation of sensory messages' Sensory Communication *(1961), 'Single units and sensation: A neuron doctrine for perceptual psychology?'* Perception *(1972) and 'Finding minimum entropy codes'* Neural Computation *(1989).*[1]

I

Alan Macfarlane (AM): It is a pleasure and honour to have a chance to talk to Horace Barlow. Horace, when and where were you born?
Horace Barlow (HB): I was born in a village in Buckinghamshire called Chesham Bois in 1921 in December.

AM: Tell me something about your ancestors, as far as you like to go, preferably not the Norman Conquest.
HB: No, I do not go back to Norman Conquest at all. But I have known about my connection with Charles Darwin,[2] who was my great-great grandfather, since I was very small because the whole Darwin family is very family conscious. Not perhaps as family conscious as some families who do go back to the Norman Conquest but equally proud.

AM: Your wife was a Darwin, was that it?
HB: No, my mother was a Darwin. She was Horace Darwin's eldest child.

AM: So she was Darwin's granddaughter?
HB: Yes. My father was the oldest son of Thomas Barlow,[3] who had the distinction of being somebody who has a disease named after him. He was also the physician to Queen Victoria's household. He was a big doctor at the time.

AM: What disease was it?
HB: What he did was he distinguished between two forms of scurvy, scurvy-rickets as it is sometimes called which is rickets due to shortage of Vitamin C whereas customarily in England the only one referred to as rickets was the one due to lack of Vitamin D. I saw a lot of him as he used to come around a lot to see his grandchildren.

This interview was conducted by Alan Macfarlane on 5 March 2012. The transcription for this interview has been provided by Radha Béteille, for this volume.
These interviews are available for viewing on:
https://www.sms.cam.ac.uk/media/1385749
https://www.youtube.com/watch?v=uBAdK0fdTdY&t
https://www.youtube.com/watch?v=Sujg21H2dzc

He had a lot of them. He was a wonderful old man. In 1929, my parents and family moved from Chesham Bois where I was born to the house that he built in 1901. The rewards for being a famous physician then were large, but he also came from a cotton family in Lancashire. I do not know that I want to go back any further than my grandparents.

AM: Did your mother talk about her grandfather? Did she know him?
HB: Oh yes, she was very enthusiastic about him and in those days thought that he was rather under-rated, though when she died she realized that perhaps he might even be overrated. So much of his reputation had increased during that time.

AM: Did you or she know Gwen Ravarat?[4]
HB: Oh yes, very closely.

AM: You knew her, did you? What was she like? She wrote *Period Piece.*
HB: She was quite a formidable character.

AM: She married an American?
HB: No, her mother was American and she was married to Jaques Raverat, who was French.

AM: Formidable, was she?
HB: Yes, yes. Every time I see *Period Piece*, I pick it up and read a paragraph or two. It's so wonderful.

AM: Let's move on then to your parents. Describe them.
HB: Yes. They were members of the intelligentsia and the upper middle classes, anxious not to be regarded as aristocratic or too posh. But an enormous number of our friends were relatives in fact, so the house was always full of other Darwins, so we saw a lot of them.

AM: Wedgewoods?
HB: And Wedgewoods, yes.

AM: Where was this? You were living in?
HB: My parents' first house was in Chesham Bois which was near Amersham.

AM: And what did your father do?
HB: He rose to be in a senior position – not the head of the Civil Service but quite high. He was quite successful. At Oxford, he read classics and then he joined the Civil Service and rose relentlessly. He was busy. He went on long after normal retirement and had many other occupations in London.

AM: And how did he influence you?
HB: He had much less influence on me than my mother. My mother took the attitude that bringing up children was not his job. I very quickly realized that my mental attitudes were not those of a classical scholar at Oxford but much closer to my mother's. She was one of William Bateson's[5] students, except she was not a student because she never came to the University. Her sister did. But she worked in Bateson's group counting seeds and offspring. She had an undergraduate friend who was interested in genetics and immediately demanded coaching in one of her papers. I think it is the only one that is quoted now that she had something to do with it. She tried to make me a botanist but I rebelled against that, the botanist side though not against the scientific side. She was very observant and would always be pointing out Darwinian details.

AM: Did you used to go on walks with her?
HB: Yes. She was a very clean gardener.

AM: Did you spend time in Grantchester because Bateson was out in Grantchester.
HB: Yes, he was in Grantchester. I do not ever remember going to the house as a child. William Bateson's wife was a friend of my mother's. I do not know if I have told this story before. When Gregory [Bateson][6] was out in Bali he met Margaret Mead.[7] Beatrice Bateson, his mother, felt she was too old to go out there and inspect her so she sent my mother out there. She flew off in an Imperial Airlines airline and we saw her off from Hendon. That must have been 1937–38. The Batesons were really …

AM: What did you make of Margaret Mead?
HB: They got on very well. I do not think they were on board instantly – I think she was not altogether convinced by Margaret

Mead is what I mean but was very impressed by her breadth of knowledge and energy and the way she expressed opinions and solutions to all sorts of problems.

AM: Did you meet her?
HB: Oh yes, she came and stayed with us many times.

AM: What was your impression?
HB: I think even more sceptical than my mother, though she was a very impressive person. Gregory too, they were a quite formidable lot.

AM: Tell me more about Gregory. You knew him from being a boy.
HB: Yes. Gregory was born in 1904 and my mother was born in 1886, so there was quite a big age difference between them. But I remember he used to come visit my mother and to stay occasionally. But I never got on close intellectual terms with Gregory even though we were to some extent interested in the same sort of thing, both in cybernetics and psychology, and his ideas were always interesting. However, my model of a scientist was taken from my mother, not from Gregory.

AM: What was your mother like as a scientist?
HB: She was interested in genetics and the paper for which she was famous was on the reproductive system in plants like cowslips. If someone had the three stages of the height of the reproductive apparatus, they could make six different types of crossbreeds, which is of some interest even now.

AM: So your mother as a scientist was an experimental scientist?
HB: She reasoned like a scientist, yes.

AM: Whereas Gregory did not?
HB: No, he was more like a bishop.

AM: Like a bishop?
HB: Well, not a bishop, like a guru. He liked to think things out for himself and that was what he taught people. He obviously was a very big influence on a lot of people. When I went to America, to Berkeley, he was in California at the time and I saw him once or twice there too.

AM: Let's come back to your early life. What is your first memory, distinctive memory, not waving leaves in the wind over your pram but a particular event?
HB: I was always interested in mechanical events and I didn't take to botany. I also did not like the natural history side of botany. It did not seem to me all that real. I was much more interested in steam engines.

AM: So as a three or four or five year old, any particular event or memory from that early childhood?
HB: I remember our first car and the moment it was driven up the drive. That must have been about 1925. My father had driven it home from having bought it somewhere in London. He was not a very good driver, always made terrible noises with the gears. My mother was much better. So I was definitely on the mechanical side.

AM: When you were a little older – that was when you were about four or five – but between five and 10, your mother was trying to get you interested in natural history but what were your particular passions or hobbies?
HB: I remember I was good at school. I was good at mathematics and not good at languages which meant they had serious worries whether I would be able to pass common entrance to get into a public school. I did succeed eventually but I think that was because Winchester had a bigger weighting factor toward mathematics than to most foreign languages.

AM: Where had you been at school before you went to Winchester?
HB: It was a day school in Chesham Bois. Then I went to a preparatory school which was run by relatives of the people who ran the day school I went to.

AM: It was a boarding school?
HB: Yes.

AM: From what age did you go there?
HB: Six, I should think.

AM: Very young.
HB: Not young for those days.

AM: Do you remember, going away to a boarding school, what your feelings were?
HB: I remember not liking it.

AM: You were homesick?
HB: Yes. I was rather a solitary boy. My mother was left partly in charge of the two youngest Cornford children because the trio – Gwen Ravarat, my mother and Frances Cornford who was a poet – as children, they grew up together and were pretty close to each other. Hugh Cornford was at the same preparatory school as I was.

AM: But you were rather lonely and kept yourself to yourself?
HB: Yes.

AM: Were you bullied?
HB: Yes, I was bullied but I survived it in the way people do. I did not mind all that much about other people's opinion of me so the bullying did not hurt as much.

AM: Were you any good at games? That often is a salvation.
HB: Yes. I was good at some sport, archery I remember. For some reason the headmaster was a keen archery fan.

AM: What was the school called?
HB: Shortenhills.

AM: Were there any teachers at the school you used to remember or not?
HB: The headmaster I remember. He was an artillery officer. He was rather keen on mathematics. I remember almost all of the other teachers. There was M. Picarde who taught us French, and Major Lownes, and there was one rather grubby person who also taught us French.

AM: When you say grubby here, are you using it in a moral sense or a physical sense?
HB: I ought to think both. But he was only there for one term.

AM: Enough to have an unpleasant memory.
HB: Yes. I went on from there to Winchester. My parents looked at all sorts of other schools as well where I could go, some of them

quite advanced. They were interested in not only all the public schools. One of my brothers went to Marlborough, the other one went to Winchester.

AM: You said you were good at maths. Were you good at maths at this age and otherwise at your preparatory school or later or both? Some mathematicians are.
HB: Some mathematicians develop early, including Petersen. I was not outstanding. I could do it and the others could not, in some cases.

AM: When you went to Winchester, were you looking to specialize in science by the second or third year?
HB: Yes, yes. There were several teachers there that I remember. Eric James. Winchester has house tutors. Some teachers were attached to each house, one from the staff who went around and taught us. Eric James, who was later the headmaster of Manchester Grammar School, was mine. Unlike many other masters at Winchester, he was liberal if not a little bit left. I remember at the time of Munich, which was just at the end of my time in Winchester, there were only four people in the whole house of about 40 pupils plus our housemaster, Eric James, who were in the least doubt that Chamberlain was a hero, whereas my parents, my mother in particular, had all sorts of connections with all sorts of people in Germany and already knew what was going on. But the others just did not seem to be aware of it at all.

AM: Do you remember it hanging over you like a cloud? My mother, who was born in 1922 – looking at her diaries, she was in terror and fear at what was happening and was so relieved when Chamberlain came back. Do you remember that?
HB: Yes, I remember the – we had several German refugees or future refugees that came to stay with us in England. In the last summer holidays in pre-war England we had two visitors from Germany. One was the son of quite a prominent Nazi and the other was Jewish. He already foresaw what was happening and got out early. One had to be rather tactful about how they interacted with each other.

AM: What else were you interested in apart from work at Winchester? Some people are interested in drama, some in....
HB: I was interested in photography.

AM: In photography?
HB: Yes. I think I was probably a rather boring schoolboy.

AM: There was music?
HB: Yes, some music. I wasn't good enough at it to be really interested.

AM: You played an instrument?
HB: I was taught piano to begin with and did it adequately. I did my grade one or two or three. I think it was three that I got up to actually in piano. But then, I really did not like it. We had an aunt who was charming. I later liked her very much. But she always used to sing folk songs and I could not stand that.

AM: You stopped playing instruments and songs, but did music....
HB: No, I went on actually. I later took up the flute and played in orchestras.

AM: Did you? Do you still play?
HB: Yes, I do, actually. Try to.

AM: Has music been important in your life?
HB: Yes, yes.

AM: When you are relaxing or – in what parts of your life does music come in? Is it just for relaxation or do you work to music or are inspired by music?
HB: I do not really like listening to music when I am not listening to it, I do not like it terribly as background. There is a quartet club in Cambridge and it is a curious organization. I still go to that quite often. I used to have my own quartet.

AM: Did you?
HB: Well, it was not my own quartet. We met more often in my house, but the first and second violins got married, and they had children.

AM: And became an orchestra. You said folk music was not your delight at that time but do you like other music? Presumably classical music is your favourite?
HB: Mainly classical music, yes.

AM: Which part of classical?
HB: The earlier part.

AM: Up to Beethoven?[8]
HB: I like Brahms,[9] and for that matter, some modern music I like. But it is never taken over my life like with some mathematicians and other academics. It has been quite important but not overwhelmingly important.

AM: You mentioned Eric James but was there any other teacher – sometimes at that age there is a teacher who really focuses you in on what you want to do with the rest of your life and inspires you.
HB: Among the hard sciences, I suppose it was Eric Lucas. He was the biology teacher but I did not do any biology because I thought I was going to come up here and do the ordinary tripos, physics and chemistry. This was just at the beginning of the war, and under strong pressure – I did not realize quite how strong then as I do now – from my mother. She said the thing for me to do was medicine, partly because early in the war they became aware of the fact that it was going to be short of doctors and exempted all medical students. I joined the biology course there right at the end and spent my first two terms here cramming organic chemistry and biology in order to get through the first MB.

AM: Why did you come to Cambridge? You had always had friends and relatives round here?
HB: My father went to Oxford so I could have gone there. It was a choice but I had reacted against classics and was more interested in science, for which Cambridge was reckoned to be better. Also, two of my brothers came to study here and one of my uncles and my grandfather.

AM: So you came to Trinity?
HB: Yes. In those days it was just a question of picking up the telephone and ringing up somebody. It so happened that one of the tutors at Trinity had been at school with my father, at Marlborough so he picked up the phone and there it was. Seems odd, does not it?

AM: Times have changed. Tell me your impression – you knew Trinity through your father and through other people. Cambridge was not much of a shock really. It was a continuation, just by another means.

HB: No. People have said it was too large but I never minded it being large. It was a bigger section of people. That was really the first time that I got selectively friendly with people. At my preparatory school, I was friendly with Hugh. He was a general practitioner in Cambridge for a long time but he was one of the most saintly people I know.

AM: Did you know his father?
HB: Only as an old gentleman in an armchair, likewise his mother. His mother was more obviously neurotic. I do not think Francis was neurotic actually. But she was, and she had a breakdown, quite serious, at the time that her eldest son, John, was killed.

AM: But he was the one who wrote *Microcosmographia Academica*?
HB: Yes, that was Francis, yes.

AM: Delightful book. Did Francis not fight in the Spanish Civil War?
HB: Yes, and so did John.

AM: John was killed there?
HB: Yes, he was killed there. Yes, but was not one of the Bruce Lockharts – was another in the....?

AM: There was one in the secret service. He was in Russia. That was earlier. But they were everywhere so probably they were there too.
HB: I have the vague recollection that he was head of the international brigade.

AM: Yes, probably. In Trinity at that time, you came up to read physics and chemistry. Was Dirac[10] there then?
HB: Yes, he was still around but I do not think anybody knew Dirac.

AM: He didn't say much, did he?
HB: No, no.

AM: Who were the others?
HB: He was in John's, Dirac. There was William Rushton,[11] there was Andrew Huxley[12] and Alan Hodgkin[13] but they were contemporaries of – my brothers were older than me and they both know Andrew and Hodgkin. There were quite a lot of physicists. I did not have supervisions from any of them because I was only doing biological subjects.

AM: Were there any teachers, any lecturers or supervisors, whom you remember distinctly as influencing you a lot?
HB: William Rushton. Were there any others? Not particularly. I do not think the chemistry teachers I had were a particularly distinguished lot actually whereas Rushton was. He was from another school that had a lot of well-known people, now one which I remember my parents considering sending me to, which is the one at Holt.

AM: Yes, Gresham's.
HB: Gresham's School.

AM: That's where a Bruce Lockhart was headmaster, Logie Bruce Lockhart.
HB: Alan Hodgkin was there and William Rushton and also some of those lovely poets. A whole string of people in the Arts world were there too and also a number of scientists. Lord Adrian – his son, Richard. Adrian was a friend of the Darwins in Cambridge, so my mother knew him pretty well. He was a contemporary of hers.

AM: You mentioned the Arts world. At school, apart from photography, you were living a fairly quiet life. When you came to university, did you smash out into new things in particular?
HB: No, not in particular.

AM: Music?
HB: Yes, I played some music.

AM: But you worked hard. This was during the war?
HB: Yes. But there were amiable times.

AM: How do you think the war most affected what you would have found at Cambridge as an undergraduate? A lot of people were away. It was subdued.
HB: Yes, but there was always plenty to do. I opted out of the Home Guard into a division of the Home Guard which used to occupy us on Saturday or Sunday as a motor cycle despatch rider. That was great fun because we were expected to get to know the highways and byways of Norfolk. You had to provide your own motor bike.

AM: You had one, did you?
HB: Well, I bought one. We went off on trips into the countryside.

AM: This was so that when the Germans infiltrated into East Anglia you would be there on the spot?
HB: The first people to be killed. But I suppose the course was compressed and people probably worked a bit harder. They kept most of it going with the supervision system, still kept going, it did not really make all that much difference.

AM: You finished in 1943–44?
HB: Well, I had a very good year actually. Two things happened. One was that I got the – the Rockefeller Foundation set up a series of medical scholarships using money that they used to have for a very extensive system of postdoctoral exchanges. But they could not get people from Europe at that stage because they were all either enlisted or conscripted or became involved in operations. So they spent the money on medical students instead and I was lucky enough to get one of those. I was supposed to go to medical school in America as soon as I left here, but I had got involved in an operational research group run by C.L. Brown on diving, at Hampstead in London. They wanted me to stay there so I delayed going to the medical school for a year. That was a very good year actually as it was my first experience of working in a research lab for any length of time, and it was also the first salary that I earned.

AM: Which year was this?
HB: It must be 1943. I was there for one full year, working on problems of divers physiologically and discovered that that was the life for me, not the cotton industry or other things that I was offered.

AM: So you spent a year there and that has taken you to 1944 and the war is still on.
HB: Then I went to America for two years as a medical student at Harvard Medical School.

AM: You were doing the MD or an MA?
HB: No, MD.

AM: I wondered how you got an MD from Harvard, so that is how.
HB: Yes. The British Government told Harvard Medical School to on no account qualify me to be a doctor because then I might have

become a doctor there. But Harvard Medical School said, 'Will the British Government dictate to us who we give our degrees to?' They went out of their way to give me one.

AM: Can you remember your feelings on going to America? That must have been a bit of a shock during the war.
HB: I felt independent. I was not independent but semi-independent and thinking and capable of acting independently by the time I went off there. America might have been thought to be a safer place than England, but actually from the company of the 20 of us that went at that time, two were torpedoed, another two got tuberculosis and died, and another two stayed there a few years. It was not a very safe option really.

AM: Did you like America? Did you think you might stay there?
HB: America evolved to suit foreigners, so we were always welcomed and I had a very good time. But they worked very hard. I was quite shocked when I came back at the end of two years as a medical student and finished off with a year at University College, to find how little work medical students did here. In many respects, the training was better than the American one in many ways because one had much more responsibility as a student and much more experience of actual clinical work. In America it was much more as it is here now. If you finished medical school, you still had lots of hospital appointments to be filled before you can be qualified. The medical training, my contemporaries who stayed behind got at least as good an education as we did.

AM: You came back just a year after the end of the war.
HB: Yes, just at the end.

AM: And you did another year. Then what did you do?
HB: Then I had to – you could get qualified without doing any resident work in a hospital. I did that and then I had to decide whether to enter the rat race and do the necessary number of residencies and house jobs. Then I thought if I was going to do research which is what I thought I wanted to do, I better have a go at it straight away, so I enrolled as a PhD student with Lord Adrian. Again that was amazingly simple once one found out how

to get hold of Adrian. He was notorious for being very difficult to get hold of. I hunted him down in the Physiology lab many times but he had a way of going down the stairs two steps at a time very close to the wall, so you could not get in his way. He escaped me once or twice like that.

AM: He was at Trinity.
HB: He was at Trinity, yes. You would not see him there. He was in the Master's Lodge.

AM: He was Master then?
HB: Yes.

AM: Tell me what he was like. He was your supervisor, was he?
HB: No, he was not. William Rushton was my supervisor except that he was not because he was director of studies. But he taught people anyway and I did quite a lot with him. In the practical classes, in the demonstrations, you get to know him like that. Unlike many supervisors, he would actually come around and talk to people, see what they understood. I really understood very well what he was like. It seemed to me like he would help during supervision, especially if one was not so good at it.

AM: But who was your PhD supervisor?
HB: That was Adrian. But it was not like a modern PhD supervisor who sees a lot of their students. Sometimes the students worked for them. But he was not like that either. He would come around once a month and just see how things were going on. He would always say interesting things, referencing, 'Have you read this?' and interesting conversation. I would not say much encouragement. As soon as I started wanting to do anything different to what he had suggested, I do not think I ever got anything except negative advice, that I should not do that. I remember that I decided that I wanted to do was the physiology of vision. I had a contemporary who was also doing that. It was at a time when you could not get money for grants very easily but there was an enormous amount of surplus war equipment for making neurophysiological apparatus. He was in London. I found out that you could get all this in government surplus stores for ridiculous prices. They sold the stuff by the pound

weight. So we got a piece of equipment that cost hundreds of pounds but it weighed ten pounds and we got it sixpence a pound.

AM: A lot of scientists in that period, people like Anthony Hewish,[14] talk about getting ex-army surplus. It had a bigger effect on Cambridge science, all these bits and pieces from the war.
HB: Yes. And anyway, there was much more equipment that was made. We had very good workshops. So even if you couldn't buy things sixpence a pound, you could get it made.

AM: You finished your PhD and then became a research fellow?
HB: Yes. At the end of my time as a PhD, I told Adrian I wanted to follow [Haldan] Keffer Hartline,[15] the first recorded work on frog's retina. I said I wanted to do that. I thought there was water to be found out there. But Adrian's advice was not to do that because I remember him saying, 'Keffer is a very clever chap'. That was his warning that I should not go into that field.

AM: Very briefly, what was your PhD?
HB: I worked first on eye movements and that was a matter of finding a good way of recording eye movements. At that time there was a theory around that eye movements were more important, people talked about scanning the images. These were the days when people were beginning to be more aware of how television worked, scanning the images, and they suggested something like that went on in the eye. If that had been the case, there was thought to be a lot more to be found out just by following eye movements. But in fact the pattern of eye movements was known since the early 1900s to be in terms of jerky movements from one fixation point to another fixation point, moving around like that. It did not encourage that theory at all. I was able to show that when a person was actually resolving something, the eye was often moving very little. Also, you can cause very much strain when the eye has only a very short period of time to see something. It sees nearly as much as it can but of course you cannot see movement because it is only one moment in time. But if you take away the actual perception of movement, the other aspects of vision progress as well as they can. It's a very short experience.

AM: Let us go on and we will come back to some of your research later. We are through to about 1949 when you ended your PhD. Is that when you came to King's?
HB: No, I failed to get a research studentship the first time I applied, and the second time I didn't.

AM: This was at Trinity?
HB: Yes, and the year I did was 1950. I was there for the next three or four years, except I included in that a year at Steve Kuffler's[16] lab in Baltimore just before he moved up to Harvard. So by the time I went to Berkeley, I had already spent a year in America actually. Well, three years altogether. It's quite a long time.

AM: So you went from Trinity to Berkeley?
HB: No, no. I went from Trinity to King's and then King's to Berkeley.

AM: Right. Were you in King's for long?
HB: I think it was eight or nine years, until in fact the assassination of Kennedy. I think I was at a feast at King's.

AM: Because everyone remembers the moment they heard the news.
HB: Yes. I was there for almost ten years and then I got an offer of a job at Berkeley which sounded fun. But there was a long delay before I got my visa and I heard that I had got my visa on the day that Kennedy was assassinated.

AM: Tell me about King's between roughly 1950 and 1960, did you have any particular friends or memories of that time?
HB: My first impression was how different from Trinity it was. The differences between colleges have got less rather than more.

AM: What were the differences?
HB: There was always a lively conversation going on somewhere at King's high table but not at Trinity. Trinity was very staid.

AM: So Dadie [Rylands],[17] you knew Dadie?
HB: Yes, but I was not one of his boys.

AM: Peter Avery you knew?
HB: No, I did not know him actually.

AM: He was probably towards the end of your time. And you knew Noel Annan?[18]

HB: I knew him, yes.

AM: Was he Provost during that time?
HB: No, he was Elector during that time. Shepherd was Provost.

AM: Shepherd was Provost. What was Shepherd like? There are many stories of Shepherd.
HB: I think he was a disappointment. He was a character but....

AM: Not very intellectually exciting.
HB: No.

AM: Who were the exciting people at that time?
HB: Well, Noel was lively, as was the pathologist, Kendal Dixon. Did you know A.V. Hill?[19]

AM: Yes. He was a Nobel Prize winner, wasn't he?
HB: Yes, he was a contemporary of Adrian's. Maurice Hill,[20] who was the son of A.V. Hill – he was a geologist.

AM: What about Donald Parry? Did you know him?
HB: Yes, yes. He was amiable and also a neurophysiologist like me.

AM: A nice man.
HB: Yes, he was. I do not think the intellectual standard was any higher in King's, but they were a less gloomy lot and more apt to talk to each other.

AM: You then went on to America. How long did you stay there?
HB: I was there for almost 10 years and I enjoyed that. I was teaching in the optometry school and it was quite easy. To fit a pair of glasses you do not need to know an awful lot about optics or physiology. So it was quite easy teaching. There was interesting research going on and I was able to do what I wanted. Berkeley is a very nice place. Do you know Berkeley?

AM: Yes, I have been to Berkeley. I liked it. It is the nicest of the universities I have been to in the States, both in location and its ethic.
HB: Yes, yes, quite. I enjoyed that.

II

AM: Were you married by that time?
HB: Yes. It was partly because my marriage was breaking up and I had some more links back in Cambridge – the funny thing about American society in general is that all my links with people here are long-lasting and deep in a sense but superficial and null in other senses whereas in America it is all changing, and the average time that a person stays in the same place is two years in America whereas here it is more like 200 – so I came back. My mother was old and dying, and I got a very nice job where your total responsibilities were really confined to writing a 200-word summary of your activities during the year. You can do other stuff you like.

AM: What was the job?
HB: Royal Society Research Professor.

AM: That is very distinguished and very heaven-on-earth. Robert Hinde[21] was one as well.
HB: Yes, he was.

AM: Wonderful job, just do what you like. It is for life, is it not?
HB: It was but I think I was one of the last one. I think it was right to change it. If you want to make the best use of people, you should not appoint them for life. People do not like that.

AM: You wrote several seminal papers around this period, according to the sources I have seen. If you had to summarize, what do you think are your most important ideas and contributions? How would you describe them?
HB: In confining it to neurophysiology, I was one of the first people to realize how much, and the type of analysis, that has to go on early in vision in order for us to be able to see as we do. This is so particularly with regard to the statistical aspects of vision, the fact that we infer so much from so little and things like that. It's because we already have learnt so much of the statistical facts, these are not things that we can say exactly what we have learnt, but we know that if we let go of a penny, it drops and all the facts about the environment and the way the world in general behaves around

us, we already have expectations formed. It is because of those expectations we can predict as well as we can. This is a principle that goes right through perception and cognition. This was not the way people thought of the brain when I started off in the 1940s, but it is very much the way in which people think of the brain now.

AM: How early in life does this appear? There used to be theories among psychologists that – there was that very famous psychologist who worked on child learning.
HB: [Jean] Piaget.[22]

AM: Piaget – that it took quite a long time before an infant could recognize patterns.
HB: Yes, you'll be talking about high-order patterns, but what I am talking to you about is things at a very elementary level. One of the first things to be brought out – how extremely sensitive the eye is in this regard, concerning the actual detection of single quanta of light, and that you can get down to actually the graininess of light itself. This limits your perception under a wide variety of circumstances. You have got to have an efficient statistical machinery there, otherwise you will not be able to do it. So some things become impossible unless you have the statistically efficient mechanisms around for determining what is associated with what. This is at very much a pre-Piagetian level. Of course, it goes on and I do not know when it stops. Well, I think I am beginning to be aware of where it stops.

AM: You are talking here about the way in which you generate a lot of information from smaller range of generation.
HB: No, you cannot generation information.

AM: You extrapolate or....
HB: You interpolate.

AM: But you also worked the other end, which is how you delete, you do not need as much information as that.
HB: When some information is permanent, you do not need to meddle with that, you use it to find out what is up to date. The redundancy reduction has often been misinterpreted. What you

are really doing is using the redundancy that you have acquired knowledge of in order to understand better what is going on now. The only definition of intelligence that makes sense to me at least is this process of using the statistical regularities that you detect in order to make inferences about the present, past, and in some cases, the future. There is a very big difference in the way of thinking about these statistics now than was prevalent when I was at school. I was not taught statistics in school, as I recall, whereas now it is an important part of everybody's education.

AM: Do you think your work – it must have directly affected a lot of work that is being done in film and computers, whereby they have worked out how to sample visual images so that you can just take cross-sections and then create a full picture but by only sampling at one twenty-fifth of a second and with recent compression, decompression, algorithms in computing, you can actually compress a huge amount of information to quite a small set of digital signals which will then decompress and human beings can see it. Do you think that your work was related to this?

HB: Yes. I was very lucky when I was both up here and later when I was working in London, knowing people like Tommy Gold and others who understood the statistical aspects of information – and Donald Mackay, who was a student of Dennis Gabor,[23] who understood all the statistical principles here. Rather few people, even now, understand information theory and what redundancy is and is not. I think it has a completely revolutionary effect on how one looks at what the brain is doing all the time. You cannot create new information. That is like spontaneous combustion, which statistical mechanics shows is impossible. But what you can do is filter. The fact about the world which is missing from earlier views is what an enormous amount of regularity there is. If you take any object or state of affairs, it is going to be almost exactly the same an instant later. All that stuff which is the same, there is no statistics to this.

AM: That is how they do the sampling of film, is it not? They just note the changes and save the changes.

HB: Yes, but the other aspect which I think is much more illuminating is if you look at language, if you take the number of words in the

English language, half a million, the fact of the matter is that you can take a random assembly of letters and it does not make a word or a sentence at all. There are so many possible arrangements of those letters which mean nothing at all, that cannot happen. The same is true of the images. If you take pure, random noise and record it on film, you cannot see anything there. It is a rainstorm, it is a snowstorm, it is completely meaningless. The only thing one sees in an image is the statistical regularities. In one single frame, you do not get very much, but in sequences of frames there is a lot that is shared between them and those are the things you see. Just as if things are arranged in a line, you detect the line immediately, but if they lack that patterned element, you do not see the line. So, what one is seeing, to say that you remove the redundancy is the reverse of the truth, the redundancy is what you see. That is what composes your images. That has been a change, you will find elements of it in the thinking of people like Ernst Mayr[24] and [Hermann] Helmholtz.[25] But there are only traces of that and it wasn't until post-war that the full significance of information theory for understanding the brain appeared, because people did not really appreciate the importance of language, the importance of all our senses. They are actually composed of redundant features, in the images that is what we use. The economy of thought was one of the earlier ways of describing this. You can find traces of it going back to the nineteenth century. But it has come to fruition with the development of cybernetic devices, which is why it has taken time. Even the people most deeply involved in it, the people first trying to do computer pattern recognition, didn't really fully appreciate the significance of this as far as perception was concerned. Having been tutored by people like Tommy Gold, Donald Mackay and others, I reckon I am one of the earliest generations of people interested in psychology who have understood the significance of this.

AM: What is your early work on frogs' brains?
HB: It was frogs' eyes. Why I wanted to go into it in the first place was the early work which Hartline had done on frog's vision had shown there were four different types of ganglion cell. We now appreciate that there are now closer to four hundred types, and they

are each telling the brain something different and something about a much larger element in what is coming into the brain. It is talking about a pattern in a particular region. The simplest example of this is motion. It is true to say that even though there were lots of illusions associated with motion which have been known since the 19th century, the fact that what is crucial in the modern understanding of the subject is that motion is one of the simplest types of patterns we detect. You cannot have motion without knowledge at two different positions in time and space. It is important that motion tells you this is changed, there has been a coordinate change of position and time of occurrence. This is one of the simplest type of things that you can use, once you have got a signal for motion, rather than a separate occurrence of things at two different times and places, then you can get joined motion at different parts. You see the eyes moving, it expands the alphabet that you can use in information theory terms. It is clear from the kind of things that we do use to express our subjective experiences are very complex elements of this type. I do not think I have done a very good job of simplifying this, making it clear. But I still think it is really rather unappreciated how significant it is. I am very lucky having lived and worked in this time because it is such an important principle which applies to so much of perception.

AM: Just two last questions I wanted to ask you particularly. One is about how you worked or work. I don't know whether you are of the Eureka brigade of scientists who remembers the moment when suddenly one of your major ideas came to you or whether it was just a gradual growth. But if you are of the Eureka type then are there particular occasions? Is it when you are walking or is it when you are listening to music, is it when you are having a shower, is it when you are drinking wine and talking to someone at High Table? What are the contexts where you find the more important ideas flow?

HB: Well, I think a lot to myself, I think while having a bath or just lying in bed, relaxing. I spend a lot of time thinking, always trying to see relations where you haven't seen them before. Like, 'good heavens, that also applies here, doesn't it', 'that's rather similar to what happens here', so you get some feels of similarity.

AM: Things happen not when you're at your lab desk but within moments of relaxation.
HB: I think so, yes. When there is novelty, when it is new, you have got to remember when it became new and that tends to be when you are not doing anything else.

AM: The other question is about religion. I often ask people about this, that when they were at their boarding schools, at Winchester, they are being confirmed. We missed that time in your life. Has religion ever been important to you? Is it important to you?
HB: No, it is not important to me, not only in the sense that I do not believe – I think that saying that you do not believe in God is a very foolish thing to say because so many people talk about it, so that cannot be right. There has got to be more to it than that. One has to respect some of the things that godly people say and even some of the things they do. So I do not have a lack of respect for it.

AM: You are not a Richard Dawkins?[26]
HB: No, no. I think that's just foolish. I wish one could make more sense of it, of that aspect of it, but I do not think the godly people have done a very good job, I concede.

AM: You were confirmed?
HB: No, I was not, no. I was never baptized or confirmed.

AM: Really? So you have never been a practitioner of religion really?
HB: No, and I do not miss it.

AM: Do you think science has disproved religion, as Dawkins argues?
HB: I think it provides some hope of acting more rationally to handle social and political problems that we have deal with, both on a person level and on a worldwide level. If you regard religion as a way of perpetuating a way of thought that might otherwise have been lost, then that is fine.

NOTES

1. Horace Barlow. 2022. Wikipedia. *https://en.wikipedia.org/wiki/Horace_Barlow*.
2. Charles Darwin, see Appendix.

3. Thomas Barlow, see Appendix.

4. Gwendolen Mary 'Gwen' Raverat née Darwin (1885–1957), was an English wood engraver who was a founder member of the Society of Wood Engravers. Her memoir *Period Piece* was published in 1952. She was the daughter of astronomer Sir George Howard Darwin and his wife, Lady Darwin (née Maud du Puy). She was the granddaughter of the naturalist Charles Darwin and a first cousin of poet Frances Cornford (née Darwin). She married the French painter Jacques Raverat in 1911. https://en.wikipedia.org/wiki/Gwen_Raverat

5. William Bateson, see Appendix.

6. Gregory Bateson, see Appendix.

7. Margaret Mead (1901–78) was an American cultural anthropologist who featured frequently as an author and speaker in the mass media during the 1960s and the 1970s. https://en.wikipedia.org/wiki/Margaret_Mead

8. Ludwig van Beethoven (1770–1827) was a German composer and pianist. Beethoven remains one of the most admired composers in the history of Western music; his works rank amongst the most performed of the classical music repertoire and span the transition from the Classical period to the Romantic era in classical music. His career has conventionally been divided into early, middle, and late periods. https://en.wikipedia.org/wiki/Ludwig_van_Beethoven

9. Johannes Brahms (1833–97) was a German composer, pianist, and conductor of the mid-Romantic period. Born in Hamburg into a Lutheran family, he spent much of his professional life in Vienna. He is sometimes grouped with Johann Sebastian Bach and Ludwig van Beethoven as one of the 'Three Bs' of music, a comment originally made by the 19th-century conductor Hans von Bülow. Brahms composed for symphony orchestra, chamber ensembles, piano, organ, voice, and chorus. A virtuoso pianist, he premiered many of his own works. https://en.wikipedia.org/wiki/Johannes_Brahms

10. Paul Dirac, see Appendix.

11. William Rushton, see Appendix.

12. Andrew Huxley, see Part 1, p. 3.

13. Alan Hodgkin, see Appendix.

14. Antony Hewish, see Appendix. See also: 'Antony Hewish' in *Creative Lives and Works: Antony Hewish, Martin Rees and Neil Turok*. 2020. In Conversation with Alan Macfarlane. Edited by Radha Béteille. London: Routledge and Social Science Press. pp. 3–46.

15. Haldan Keffer Hartline, see Appendix.

16. Steve Kuffler, see Appendix.

17. George Humphrey Wolferstan Rylands, CH, CBE (1902–99) known as Dadie Rylands, was a British literary scholar and theatre director. https://en.wikipedia.org/wiki/Dadie_Rylands

18. Noel Gilroy Annan, Baron Annan, OBE (1916–2000) was a British military intelligence officer, author, and academic. During his military career, he rose to the rank of colonel and was appointed to the Order of the British Empire as an Officer (OBE). He was provost of King's College, Cambridge, 1956–66, provost of University College London, 1966–78, vice-chancellor of the University of London, and a member of the House of Lords. https://en.wikipedia.org/wiki/Noel_Annan,_Baron_Annan

19. A.V. Hill, see Appendix.

20. Maurice Neville Hill, FRS (1919–66) was a British marine geophysicist. Hill was the son of Nobel Prize–winning physiologist Archibald Vivian Hill and his wife Margaret Hill, the daughter of John Neville Keynes and sister of John Maynard Keynes. https://en.wikipedia.org/wiki/Maurice_Hill_(geophysicist)

21. Robert Hinde, see Appendix.

22. Jean Piaget, see Appendix.

23. Dennis Gabor, see Appendix.

24. Ernst Mayr, see Appendix.

25. Hermann Helmholtz, see Appendix.

26. Richard Dawkins, FRS, FRSL (b. 1941) is a British evolutionary biologist and author. He is an emeritus fellow of New College, Oxford, and was Professor for Public Understanding of Science in the University of Oxford from 1995 to 2008. An atheist, he is well known for his criticism of creationism and intelligent design. https://en.wikipedia.org/wiki/Richard_Dawkins

Appendix
Biographical Information
(in alphabetical order)

EDGAR DOUGLAS ADRIAN, 1ST BARON ADRIAN (1889–1977)

Adrian was a British electrophysiologist who with Sir Charles Sherrington won the Nobel Prize for Physiology or Medicine in 1932 for discoveries regarding the nerve cell. Adrian graduated in medicine in 1915 from Trinity College, Cambridge. After medical service during World War I, he spent the greater part of his professional life at Cambridge in research and teaching, and as master of Trinity College (1961–65) and chancellor of the University (1968–75).

Adrian researched nerve impulses from sense organs, amplifying variations in electrical potential and recording smaller potential changes than had been detectable previously. Later he recorded nerve impulses from single sensory endings and motor nerve fibres, measurements contributing to a better understanding of the physical basis of sensation and the mechanism of muscular control. After 1934 Adrian studied the electrical activity of the brain; his work on the variations and abnormalities of the changes known as the Berger rhythm opened new fields of investigation in epilepsy and in the location of cerebral lesions. https://www.britannica.com/biography/Edgar-Douglas-Adrian-1st-Baron-Adrian-of-Cambridge

MATTHEW ARNOLD (1822–88)

Arnold was an English poet and cultural critic who worked as an inspector of schools. He was the son of Thomas Arnold, the celebrated headmaster of Rugby School, and brother to both Tom Arnold, literary professor, and William Delafield Arnold, novelist and colonial administrator. Matthew Arnold has been characterised as a sage writer, a type of writer who chastises and instructs the reader on contemporary social issues. He was also an inspector of schools for thirty-five years, and supported the concept of state-regulated secondary education. https://en.wikipedia.org/wiki/Matthew_Arnold

THOMAS ARNOLD (1795–1842)

Arnold was an English educator and historian. Arnold was an early supporter of the Broad Church Anglican movement. He was the headmaster of Rugby School from 1828 to 1841, where he introduced a number of reforms that were widely copied by other prestigious public schools. His reforms redefined standards of masculinity and achievement. https://en.wikipedia.org/wiki/Thomas_Arnold

EMMA NORA BARLOW, LADY BARLOW NÉE DARWIN (1885–1989)

Barlow was a British botanist and geneticist. The granddaughter of the British naturalist Charles Darwin, Barlow began her academic career studying botany at Cambridge under Frederick Blackman, and continued her studies in the new field of genetics under William Bateson from 1904 to 1906. Her primary research focus when working with Bateson was the phenomenon of herostylism within the primrose family. In later life she was one of the first Darwinian scholars, and founder of the Darwin Industry of scholarly research into her grandfather's life and discoveries. https://en.wikipedia.org/wiki/Nora_Barlow

SIR THOMAS BARLOW, 1ST BARONET, KCVO, FRS, FRCP (1845–1945)

Barlow was a British royal physician, known for his research on infantile scurvy. He studied as an undergraduate at Manchester and London. University College London (UCL) Bachelor of Medicine

(BM) in 1873 and Doctor of Medicine (MD) 1874. He became a registrar at Great Ormond Street Hospital, and later a physician and in 1899 a consultant. He was professor at the UCL from 1895 to 1907, initially of paediatrics and later of clinical medicine. In 1883, he showed that infantile scurvy was identical with adult scurvy. Barlow's disease – infantile scurvy – is named after him.

He was Royal Physician to Queen Victoria and attended her on her death, and to Edward VII and King George V. He was knighted as a Knight Commander of the Royal Victorian Order in March 1901, and in February 1902 he was created a Baronet, of Wimpole Street in St Marylebone in the County of London. He was President of the Royal College of Physicians from 1910–14 and delivered their Harveian Oration in 1916 on the subject of Harvey, The Man and the Physician. He was elected a Foreign Honorary Member of the American Academy of Arts and Sciences in 1918. https://en.wikipedia.org/wiki/Sir_Thomas_Barlow,_1st_Baronet

GREGORY BATESON (1904–80)

Bateson was an English anthropologist, social scientist, linguist, visual anthropologist, semiotician, and cyberneticist whose work intersected that of many other fields. His writings include *Steps to an Ecology of Mind* (1972) and *Mind and Nature* (1979).

Bateson's interest in systems theory forms a thread running through his work. He was one of the original members of the core group of the Macy conferences in Cybernetics (1941–60), and the later set on Group Processes (1954–60), where he represented the social and behavioural sciences. He was interested in the relationship of these fields to epistemology. His association with the editor and author Stewart Brand helped widen his influence. https://en.wikipedia.org/wiki/Gregory_Bateson

WILLIAM BATESON (1861–1926)

Bateson was an English biologist who was the first person to use the term genetics to describe the study of heredity, and the chief populariser of the ideas of Gregor Mendel following their rediscovery in 1900 by Hugo de Vries and Carl Correns. His 1894 book *Materials for the Study of Variation* was one of the earliest formulations of the new approach to genetics. https://en.wikipedia.org/wiki/William_Bateson

PATRICK MAYNARD STUART BLACKETT, BARON BLACKETT, OM, CH, FRS (1897–1974)

Blackett was a British experimental physicist known for his work on cloud chambers, cosmic rays, and paleomagnetism, winning the Nobel Prize for Physics in 1948. In 1925 he became the first person to prove that radioactivity could cause the nuclear transmutation of one chemical element to another. He also made a major contribution in World War II advising on military strategy and developing operational research. His left-wing views saw an outlet in third world development and in influencing policy in the Labour Government of the 1960s. https://en.wikipedia.org/wiki/Patrick_Blackett

EDUARDO BRAUN-MENÉNDEZ (1903–59)

Braun-Menéndez was a noted Argentine physiologist. He studied at the Faculty of Medical Sciences at the University of Buenos Aires, choosing cardiovascular medicine and physiology as his specialties. His doctoral thesis dealt with the relationship between the pituitary gland and diencephalon to blood pressure, and was developed at the Institute of Physiology under the supervision of Noble laureate Dr. Bernardo Houssay, in 1934. After receiving his doctoral degree, he went to England to study at the University College London, where he investigated the metabolism of the heart.

Braun-Menéndez also helped to create the important scientific journal *Ciencia y Investigación* which published its first issue in 1945. It was directed by him until 1959, the year that he died. Another of his initiatives included the *Acta Physiologica Latinoamericana*, a publication written in multiple languages for the publication of the work of Latin American physiologists. https://en.wikipedia.org/wiki/Eduardo_Braun-Menéndez

CARLOS JUSTINIANO RIBEIRO CHAGAS (1879–1934)

Chagas was a Brazilian sanitary physician, scientist, and bacteriologist who worked as a clinician and researcher. He discovered Chagas disease, also called *American trypanosomiasis*, in 1909, while he was working at the Oswaldo Cruz Institute in Rio de Janeiro.

Chagas's work holds a unique place in the history of medicine. Working in primitive conditions, Chagas described in detail a previously-unknown infectious disease, its pathogen, vector (*Triatominae*), host, clinical manifestations, and epidemiology. Chagas was also the first to discover and illustrate the parasitic fungal genus *Pneumocystis*, which later became infamous for being linked to pneumocystis pneumonia in AIDS patients. https://en.wikipedia.org/wiki/Carlos_Chagas

KENNETH S. 'KACY' COLE'S (1900–84)

Kenneth Cole is considered the father of biophysics. He was the world's leading authority on the biophysics of the nervous system. His studies of the transmission of nerve impulses, especially those of electrical resistance in the nerve cells of squid, are credited with leading to the rapid advances in neuro-physiology seen during the 1930s and 1940s. Cole created the innovative technique known as voltage clamping that helped set the stage for the formulation of the 'sodium theory' by Alan L. Hodgkin and Andrew F. Huxley, for which they were awarded the Nobel Prize in 1947. He was elected to the National Academy of Sciences in 1956 and received the National Medal of Science in 1967. https://bit.ly/3sj1mZm

CHARLES ROBERT DARWIN, FRS, FRGS, FLS, FZS (1809–82)

Darwin was an English naturalist, geologist and biologist, best known for his contributions to evolutionary biology. His proposition that all species of life have descended from common ancestors is now widely accepted and considered a fundamental concept in science.

In a joint publication with Alfred Russel Wallace, he introduced his scientific theory that this branching pattern of evolution resulted from a process that he called natural selection, in which the struggle for existence has a similar effect to the artificial selection involved in selective breeding. Darwin has been described as one of the most influential figures in human history, and he was honoured by burial in Westminster Abbey.

Darwin published his theory of evolution with compelling evidence in his 1859 book *On the Origin of Species*. By the 1870s, the scientific community and a majority of the educated public had accepted evolution as a fact. However, many favoured competing explanations that gave only a minor role to natural selection, and it was not until the emergence of the modern evolutionary synthesis from the 1930s to the 1950s that a broad consensus developed in which natural selection was the basic mechanism of evolution. Darwin's scientific discovery is the unifying theory of the life sciences, explaining the diversity of life. https://en.wikipedia.org/wiki/Charles_Darwin

PAUL ADRIEN MAURICE DIRAC, OM, FRS (1902–84)

Dirac was a British theoretical physicist who is regarded as one of the most significant physicists of the 20th century. Dirac made fundamental contributions to the early development of both quantum mechanics and quantum electrodynamics. Among other discoveries, he formulated the Dirac equation which describes the behaviour of fermions and predicted the existence of antimatter. Dirac shared the 1933 Nobel Prize in Physics with Erwin Schrödinger 'for the discovery of new productive forms of atomic theory'. He also made significant contributions to the reconciliation of general relativity with quantum mechanics. https://en.wikipedia.org/wiki/Paul_Dirac

DENNIS GABOR, CBE, FRS (1900–79)

Gabor was a Hungarian-British electrical engineer and physicist, most notable for inventing holography, for which he later received the 1971 Nobel Prize in Physics. He obtained British citizenship in 1934, and spent most of his life in England. https://en.wikipedia.org/wiki/Dennis_Gabor

HALDAN KEFFER HARTLINE (1903–83)

Hartline was an American physiologist who was a cowinner (with George Wald and Ragnar Granit) of the 1967 Nobel Prize for Physiology or Medicine for his work in analysing the neurophysiological mechanisms of vision. Hartline began his study of retinal electrophysiology as a National Research Council Fellow at Johns Hopkins University, Baltimore, receiving his MD in 1927. After attending the universities of Leipzig and Munich as an Eldridge Johnson traveling research scholar, he became professor of biophysics and chairman of the department at Johns Hopkins in 1949. He joined the staff of Rockefeller University, New York City, in 1953 as professor of neurophysiology. https://www.britannica.com/biography/Haldan-Keffer-Hartline

HERMANN LUDWIG FERDINAND VON HELMHOLTZ (1821–94)

Helmholtz was a German physicist and physician who made significant contributions in several scientific fields. The largest German association of research institutions, the Helmholtz Association, is named after him. In physiology and psychology, he is known for his mathematics of the eye, theories of vision, ideas on the visual perception of space, colour vision research, and on the sensation of tone, perception of sound, and empiricism in the physiology of perception. In physics, he is known for his theories on the conservation of energy, work in electrodynamics, chemical thermodynamics, and on a mechanical foundation of thermodynamics. As a philosopher, he is known for his philosophy of science, ideas on the relation between the laws of perception and the laws of nature, the science of aesthetics, and ideas on the civilizing power of science. https://en.wikipedia.org/wiki/Hermann_von_Helmholtz

ANTONY HEWISH, FRS, FINSTP (1924–2021)

Hewish was a British radio astronomer who won the Nobel Prize for Physics in 1974 (together with fellow radio-astronomer Martin Ryle) for his role in the discovery of pulsars. He was also awarded the Eddington Medal of the Royal Astronomical Society in 1969. Hewish proposed the construction of a large phased array radio telescope, which could be used to perform a survey at high time resolution, primarily for studying interplanetary scintillation. In 1965 he secured funding to construct his design, the Interplanetary Scintillation Array, at the Mullard Radio Astronomy Observatory (MRAO) outside Cambridge. https://en.wikipedia.org/wiki/Antony_Hewish

ROBERT AUBREY HINDE, CBE, FRS, FBA (1923–2016)

Hinde was a British zoologist, ethologist and psychologist. He served as the Emeritus Royal Society Research Professor of Zoology at the University of Cambridge. Hinde is best known for his ethological contributions to the fields of animal behaviour and developmental psychology. https://en.wikipedia.org/wiki/Robert_Hinde

ALAN LLOYD HODGKIN (1914–98)

Hodgkin was educated at the Downs School, Malvern (1923–27), Greshams School, Holt (1927–32), and Trinity College, Cambridge (1932–36). His grandfather, Thomas Hodgkin, and uncle, Robin Hodgkin, were historians and to begin with Alan hesitated between history and science. However, he was strongly interested in natural history and that decided him to take up biology and chemistry. After he had become a scholar at Trinity, his future zoology teacher, Carl Pantin, advised him to learn as much mathematics and physics as he could. This was good, if painful, advice which has kept him busy ever since. As an undergraduate he started some rather amateur experiments on frog nerve and continued this line for several years, first as a research scholar and later as a fellow of Trinity. At that period the high table of Trinity included an astonishing array of scientific talent, and Hodgkin found it inspiring if sometimes daunting to meet people like J.J. Thomson, Rutherford, Aston, Eddington, Hopkins, G.H. Hardy and Adrian. In the Physiological Laboratory he learnt about cable-theory from Rushton and about amplifiers from Matthews, Grey Walter and Rawdon-Smith. https://www.nobelprize.org/prizes/medicine/1963/hodgkin/biographical/

ALDOUS HUXLEY (1894–1963)

Aldous Huxley was an English writer and philosopher. He wrote nearly 50 books – both novels and non-fiction works – as well as wide-ranging essays, narratives, and poems. Born into the prominent Huxley family, he graduated from Balliol College, Oxford, with an undergraduate degree in English literature. Early in his career, he published short stories and poetry and edited the literary magazine Oxford Poetry, before going on to publish travel writing, satire, and screenplays. He spent the latter part of his life in the United States, living in Los Angeles from 1937 until his death. By the end of his life, Huxley was widely acknowledged as one of the foremost intellectuals of his time. He was nominated for the Nobel Prize in Literature nine times and was elected Companion of Literature by the Royal Society of Literature in 1962. Huxley was a pacifist. He grew interested in philosophical mysticism and universalism, addressing these subjects with works such as *The Perennial Philosophy* (1945) – which illustrates commonalities between Western and Eastern mysticism – and *The Doors of Perception* (1954) – which interprets his own psychedelic experience with mescaline. In his most famous novel *Brave New World* (1932) and his final novel *Island* (1962), he presented his vision of dystopia and utopia, respectively. https://en.wikipedia.org/wiki/Aldous_Huxley

DAVID BRUCE HUXLEY (1915–92)

David Bruce Huxley was a financier and lawyer. He served in World War II in Africa and Iraq reaching the rank of Brigade

Major in the British Army. He became the youngest Queen's Counsel (QC) in the British Empire. In Bermuda in the 1940s and 1950s he was Solicitor General, Attorney General, and acting Chief Justice of the Supreme Court. He compiled and revised many of the laws of Bermuda. He married twice and had five children by his first wife, Anne Remsen Schenck. His daughter Angela Mary Bruce Huxley (born 1939) married George Pember Darwin in 1964, and his son Michael Remsen Huxley (born 1940) became curator of science at the Smithsonian Institution. In retirement, David and his second wife, Ouida (who was raised by her aunt Ouida Rathbone, married to the actor Basil Rathbone) lived in Wansford-in-England, Cambridgeshire, where he served as churchwarden. https://en.wikipedia.org/wiki/Huxley_family#David_Bruce_Huxley

SIR JULIAN SORELL HUXLEY, FRS (1887-1975)

Julian Huxley was an English evolutionary biologist, eugenicist, and internationalist. He was a proponent of natural selection, and a leading figure in the mid-twentieth century modern synthesis. He was secretary of the Zoological Society of London (1935–42), the first Director of UNESCO, a founding member of the World Wildlife Fund, the president of the British Eugenics Society (1959–1962), and the first President of the British Humanist Association. Huxley was well known for his presentation of science in books and articles, and on radio and television. He directed an Oscar-winning wildlife film. He was awarded UNESCO's Kalinga Prize for the popularisation of science in 1953, the Darwin Medal of the Royal Society in 1956, and the Darwin–Wallace Medal of the Linnaean Society in 1958. He was also knighted in that same year, 1958, a hundred years after Charles Darwin and Alfred Russel Wallace announced the theory of evolution by natural selection. In 1959 he received a Special Award of the Lasker Foundation in the category Planned Parenthood – World Population. https://en.wikipedia.org/wiki/Julian_Huxley

THOMAS HENRY HUXLEY, PC, FRS, HONFRSE, FLS (1825–95)

Thomas Huxley was an English biologist and anthropologist specialising in comparative anatomy. He is known as 'Darwin's Bulldog' for his advocacy of Charles Darwin's theory of evolution. The stories regarding Huxley's famous debate in 1860 with Samuel Wilberforce were a key moment in the wider acceptance of evolution and in his own career, although historians think that the surviving story of the debate is a later fabrication. Huxley had been planning to leave Oxford on the previous day, but, after an encounter with Robert Chambers, the author of *Vestiges*, he changed his mind and decided to join the debate. Wilberforce was coached by Richard Owen, against whom Huxley also debated about whether humans were closely related to apes. Huxley was slow to accept some of Darwin's ideas, such as gradualism, and was undecided about natural selection, but despite this he was wholehearted in his public support of Darwin. Instrumental in developing scientific education in Britain, he fought against the more extreme versions of religious tradition. Originally coining the term in 1869, Huxley elaborated on 'agnosticism' in 1889 to frame the nature of claims in terms of what is knowable and what is not. https://en.wikipedia.org/wiki/Thomas_Henry_Huxley

STEPHEN WILLIAM KUFFLER, FORMEMRS (1913–80)

Kuffler was a pre-eminent Hungarian-American neurophysiologist. He is often referred to as the 'Father of Modern Neuroscience'. Kuffler, alongside noted Nobel Laureates Sir John Eccles and Sir Bernard Katz gave research lectures at the University of Sydney, strongly influencing its intellectual environment while working at Sydney Hospital. He founded the Harvard Neurobiology department in 1966, and made numerous seminal contributions to our understanding of vision, neural coding, and the neural implementation of behaviour. He is known for his research on neuromuscular junctions in frogs, presynaptic inhibition, and the neurotransmitter GABA. In 1972, he was awarded the Louisa Gross Horwitz Prize from Columbia University. https://en.wikipedia.org/wiki/Stephen_Kuffler

SIR THOMAS LEWIS, CBE, FRS, FRCP (1881–1945)

Lewis was a British cardiologist (although he personally disliked the term, preferring cardiovascular disease specialist) He coined the term 'clinical science'. https://en.wikipedia.org/wiki/Thomas_Lewis_(cardiologist)

ERNST WALTER MAYR (1904–2005)

Mayr was one of the 20th century's leading evolutionary biologists. He was also a renowned taxonomist, tropical explorer, ornithologist, philosopher of biology, and historian of science. His work contributed to the conceptual revolution that led to the modern evolutionary synthesis of Mendelian genetics, systematics, and Darwinian evolution, and to the development of the biological species concept.

Although Charles Darwin and others posited that multiple species could evolve from a single common ancestor, the mechanism by which this occurred was not understood, creating the species problem. Ernst Mayr approached the problem with a new definition for species. In his book *Systematics and the Origin of Species* (1942) he wrote that a species is not just a group of morphologically similar individuals, but a group that can breed only among themselves, excluding all others. When populations within a species become isolated by geography, feeding strategy, mate choice, or other means, they may start to differ from other populations through genetic drift and natural selection, and over time may evolve into new species. The most significant and rapid genetic reorganization occurs in extremely small populations that have been isolated (as on islands).

His theory of peripatric speciation (a more precise form of allopatric speciation which he advanced), based on his work on birds, is still considered a leading mode of speciation, and was the theoretical underpinning for the theory of punctuated equilibrium, proposed by Niles Eldredge and Stephen Jay Gould. Mayr is sometimes credited with inventing modern philosophy of biology, particularly the part related to evolutionary biology, which he distinguished from physics due to its introduction of (natural) history into science. https://en.wikipedia.org/wiki/Ernst_Mayr

ROBERT ALEXANDER MCCANCE, CBE, FRS (1898–1993)

McCance was a paediatrician, physiologist, biochemist and nutritionist and was the first Professor of Experimental Medicine at the University of Cambridge. Remaining at King's College in the 1930s, McCance conducted research on himself, studying the physiological effects of salt. He would use volunteers who would eat a salt free diet, and lie sweating under heat lamps for two hours a day for 10 days, to remove salt from the body, and become salt deficient. This would lead to suffering in the volunteers including cramps and shortness of breath as well as anorexia and nausea. This led to further research which established that infants with salt deficiency, excreted little salt. A later line of research, initiated with a patient who had polycythaemia rubra vera led to research which established that the amount of iron in the body is regulated, not by intestinal excretion as had been taught, but by controlled absorption. https://en.wikipedia.org/wiki/Robert_McCance

ALEXANDER LUDWIG VON MURALT (1903–90)

Muralt was a Swiss physicist and doctor After graduating from the canton school in Zurich, von Muralt began studying at the University of Zurich in 1921. He studied physics with Erwin Schrödinger and at the same time took the basic scientific subjects of medicine. In 1926 he received the second medical preparatory course and a year later in 1927 the doctoral examination in physics followed, which he completed summa cum laude. He then went in 1928 to the Institute for Physical Chemistry at Harvard University in Boston, where he was under the direction of Edwin Cohn and together with the doctoral student John T. Edsall examined the physico-optical properties of the later discovered muscle protein actomyosin. Von Muralt had a lifelong friendship with Edsall. The 27-year-old von Muralt turned down an offered professorship in biophysics at Harvard University in order to complete the clinical part of medical studies at Heidelberg University. https://de.wikipedia.org/wiki/Alexander_von_Muralt

JEAN WILLIAM FRITZ PIAGET (1896–1980)

Piaget was a Swiss psychologist known for his work on child development. Piaget's theory of cognitive development and epistemological view are together called 'genetic epistemology'. By the end of the 20th century, Piaget was second only to B.F. Skinner as the most-cited psychologist of that era. https://en.wikipedia.org/wiki/Jean_Piaget

MARTIN JOHN SPENCER RUDWICK (B. 1932)

Rudwick is a British geologist, historian, and academic. Rudwick is an emeritus professor of History at the University of California, San Diego and an affiliated research scholar at Cambridge University's Department of History and Philosophy of Science. His principal field of study is the history of the earth sciences; his work has been described as the 'definitive histories of the pre-Darwinian earth sciences'. Rudwick was an early scholar to critique the conflict thesis regarding religion and science. https://en.wikipedia.org/wiki/Martin_J._S._Rudwick

SIR JAMES COCHRAN STEVENSON RUNCIMAN, CH, FBA (1903–2000)

Runicman was an English historian best known for his three-volume A History of the Crusades (1951–54). His three-volume history has had a profound impact on common conceptions of the Crusades, primarily portraying the Crusaders negatively and the Muslims favourably. Runciman was a strong admirer of the Byzantine Empire, and consequently held a bias against the Crusaders for the Fourth Crusade evident in his work. While praised by older crusade historians as a storyteller and prose stylist, he is viewed as biased by some contemporary historians. https://en.wikipedia.org/wiki/Steven_Runciman

WILLIAM ALBERT HUGH RUSHTON, FRS (1901–80)

Rushton was professor of Physiology at Trinity College, Cambridge. His main interest lay in colour vision and his Principle of Univariance is of seminal importance in the study of perception. https://en.wikipedia.org/wiki/W._A._H._Rushton

JOHN RYLE (1889–1950)

Ryle was a British physician and epidemiologist. He was born the son of Brighton doctor R.J. Ryle and brother of the Oxford philosopher Gilbert Ryle. He was educated at Brighton College and Guy's Hospital where he qualified in 1913. He served in the military during World War I and afterwards qualified MD at the University of London. After teaching at Guy's Hospital, he was appointed in 1935 Regius Professor of Physic [not Physics; 'Physic' here is an archaic term for Medicine] at the University of Cambridge. In 1943 he was appointed chair of the newly created Institute of Social Medicine at the University of Oxford, initiating the academic discipline of Social Medicine (Epidemiology). https://en.wikipedia.org/wiki/John_Ryle_(physician)

ROBERT STÄMPFLI (1914–2002)

Stämpfli was a Swiss physiologist and membrane researcher who, together with Sir Andrew Fielding Huxley, succeeded in first describing the saltatory stimulus propagation on myelinated nerve fibres. As the spiritus rector of the Collaborative Research Center for Membrane Research, he succeeded in establishing the Physiological Institute at Saarland University as a world-renowned centre for electrophysiology and membrane biology. https://bit.ly/2Xn34gW

ELSIE WIDDOWSON, CH, CBE, FRS (1906–2000)

Widdowson was a British dietitian and nutritionist. She and Dr Robert McCance, a pediatrician, physiologist, biochemist, and nutritionist, were responsible for overseeing the government-mandated addition of vitamins to food and wartime rationing in Britain during World War II. https://en.wikipedia.org/wiki/Elsie_Widdowson

JOHN ZACHARY YOUNG, FRS (1907–97)

Young was an English zoologist and neurophysiologist, described as 'one of the most influential biologists of the 20th century'. Most of Young's scientific research was on the nervous system. He discovered the squid giant axon and the corresponding squid giant synapse. His work in the 1930s on signal transmission in, and the fibre structure of, nerves inspired the work of Sir Andrew Huxley and Sir Alan Hodgkin for which they received a Nobel Prize.

During World War II, responding to the large number of nerve injuries sustained by soldiers in combat and drawing on his work in comparative anatomy and the regrowth of damaged nerves in squids and octopuses, Young set up a unit at the University of Oxford to study nerve regeneration in mammals. His team investigated the biochemical conditions which control nerve fibre growth and also sought ways to accelerate the repair of peripheral nerves severed by injury. Working with Peter Medawar, Young found a way to re-join small peripheral nerves using a 'glue' of plasma. This method was eventually modified and used in surgery. After WWII, Young's research interests turned to investigating the central nervous system and the functions of the brain. He discussed and corresponded with the mathematician Alan Turing on brain cells, memory, pattern recognition, and embryology, from 1949. https://en.wikipedia.org/wiki/John_Zachary_Young

Social Science Press is grateful and therefore acknowledges the various websites from which information was drawn to create the endnotes and Appendix. All the websites have been duly cited, but a special thanks is owed to Wikipedia and Britannica.

For Product Safety Concerns and Information please contact our EU
representative GPSR@taylorandfrancis.com
Taylor & Francis Verlag GmbH, Kaufingerstraße 24, 80331 München, Germany

www.ingramcontent.com/pod-product-compliance
Ingram Content Group UK Ltd.
Pitfield, Milton Keynes, MK11 3LW, UK
UKHW020130200926
475702UK00005B/677
* 9 7 8 1 0 3 2 4 0 5 0 6 3 *